The Python Audio Cookbook

The Python Audio Cookbook offers an introduction to Python for sound and multimedia applications, with chapters that cover writing your first Python programs, controlling Pyo with physical computing, and writing your own GUI, among many other topics.

Guiding the reader through a variety of audio synthesis techniques, the book empowers readers to combine their projects with popular platforms, from the Arduino to Twitter, and state-of-the-art practices such as AI. *The Python Audio Cookbook* balances accessible explanations for theoretical concepts, including Python syntax, audio processing and machine learning, with practical applications.

This book is an essential introductory guide to Python for sound and multimedia practitioners, as well as programmers interested in audio applications.

Alexandros Drymonitis is a sound and new media artist. He has a PhD in the creation of musical works with the programming language Python. His artistic practice focuses on new techniques utilising new media such as computer programming, live coding, AI, or even older practices, like modular synthesis.

The Python Audio Cookbook

Recipes for Audio Scripting with Python

Alexandros Drymonitis

LONDON AND NEW YORK

Designed cover image: © pressureUA / Getty images

First published 2024
by Routledge
4 Park Square, Milton Park, Abingdon, Oxon OX14 4RN

and by Routledge
605 Third Avenue, New York, NY 10158

Routledge is an imprint of the Taylor & Francis Group, an informa business

© 2024 Alexandros Drymonitis

British Library Cataloguing-in-Publication Data
A catalogue record for this book is available from the British Library

Library of Congress Cataloging-in-Publication Data
Names: Drymonitis, Alexandros, author.
Title: The Python audio cookbook : recipes for audio scripting with Python / Alexandros Drymonitis.
Description: Abingdon, Oxon ; New York : Routledge, 2023. | Includes bibliographical references and index. | Identifiers: LCCN 2023030282 (print) | LCCN 2023030283 (ebook) | ISBN 9781032480114 (paperback) | ISBN 9781032480145 (hardback) | ISBN 9781003386964 (ebook)
Subjects: LCSH: Python (Computer program language) | Computer sound processing. | Software synthesizers.
Classification: LCC ML74.4.P97 D78 2023 (print) | LCC ML74.4.P97 (ebook) | DDC 780.285—dc23/eng/20230829
LC record available at https://lccn.loc.gov/2023030282
LC ebook record available at https://lccn.loc.gov/2023030283

ISBN: 978-1-032-48014-5 (hbk)
ISBN: 978-1-032-48011-4 (pbk)
ISBN: 978-1-003-38696-4 (ebk)

DOI: 10.4324/9781003386964

Typeset in Times New Roman
by Apex CoVantage, LLC

To Katia and Alina

Contents

Acknowledgements

I would like to thank Olivier Bélanger, the developer of the Pyo module for DSP in Python, as this book is based entirely on this module. Apart from the development of Pyo, I would also like to thank him for supporting its community and updating this module with new features. Pyo is a free and open-source module that Olivier is developing in his free time. I hope this book will help to expand the Pyo community, as this software deserves more attention than it currently gets.

Hannah Rowe and Emily Tagg from Routledge have been very helpful and friendly and replied promptly to any queries I had during the writing of this book.

I would like to thank Anton Mikhailov for providing code for the drum machine project in Chapter 3.

Louizos, Terry, and Yannis helped me out by trusting me with their computers so I could test things out.

Glossary

A

ADSR Acronym for Attack – Decay – Sustain – Release, the four stages of a standard amplitude envelope that simulates the amplitude envelope of acoustic instruments

API Acronym for Application Programming Interface. An API is a set of rules that enables different applications to communicate with each other.

Additive Synthesis Computer music synthesis technique, where simple tones with no or a few harmonics, like sine waves, are added together, to create a more complex waveform.

Aliasing An effect that occurs in digital audio, when a frequency higher than half the sampling rate (called the Nyquist frequency) is attempted to be produced by an oscillator. Since each oscillation cycle requires at least two points to be sampled, the highest frequency that can be produced is half the sampling rate. When a frequency higher than that is attempted to be produced, the computer samples less frequently than required, resulting in a tone being produced with a pitch lower than the Nyquist frequency, as much as the attempted frequency is higher than that.

Amplitude Modulation (AM) Computer music synthesis technique where an oscillator of (usually) a low frequency modulates the amplitude of a sound source.

Arduino Electronics prototyping board and programming language for physical computing. Enables the user to utilise sensors in their projects, but also to control elements like motors, lights, solenoids and others, through computer software.

Asynchronous Type of Granular Synthesis, where grains are being triggered sparsely instead of at regular intervals, resulting in a more spread sound.

B

Baud The speed at which data is sent over a communication channel. Named after its inventor, Émile Baudot.

Backend Part of software not visible to the user, usually referred to as the data access layer. Opposite of frontend, usually called the presentation layer, which is the part of the software visible to the user.

Band-limited Term referring to oscillators that have a fixed number of harmonics, usually created with Additive Synthesis.

Bandpass Type of filter that lets a band around its cutoff frequency to pass, while frequencies on either side of its band are attenuated. The width of the band is controlled by the Q.

Bipolar Refers to oscillators whose amplitude alternates between positive and negative. These oscillators are used for audio, whereas unipolar oscillators are used for control.

Bit depth   The number of bits used to represent amplitude samples of audio. The more bits used, the greater the dynamic range of an audio signal. The bit depth also determines the signal-to-noise ratio (SNR) in a signal.

Boolean   Data type that is common among programming languages, named after the mathematician George Boole. A Boolean value can be either True or False.

Buffer size   The number of samples stored in an audio system, before these are processed, also referred to as the block size.

C

C   A general-purpose programming language developed by Dennis Ritchie at Bell Labs, in the early 1970s. It is used by many operating systems and software engineers. It is a common language among audio processing environments.

C++   A general-purpose programming language developed in the 1980s by Bjarne Stroustrup, as an extension to C.

Carrier   The sounding oscillator in various synthesis techniques like Frequency Modulation, Ring Modulation, Amplitude Modulation, and Phase Modulation.

Casting   Technique that changes the type of a datum, used in languages like C and C++.

Class   Abstraction of code that creates a structure with variables and methods, inherent to instances of the class that are created when a program runs.

Compiler   A computer program that translates computer code written in one programming language, to another, low-level programming language, referred to as machine code. The latter is used to create an executable program.

Cross-synthesis   Synthesis technique that mixes the amplitudes of one sound source, with the frequencies of another.

D

DFT   Discrete Fourier Transform. An algorithm that outputs a discrete-frequency spectrum from a short sound segment. Named after the French mathematician and physicist Jean-Baptiste Joseph Fourier

Directory   A place in a computer file system with a tree-structure. A directory is similar to, and often inter-changeable with the concept of the folder, which is used to organise the files in a computer system. A directory might contain files or other directories. The root directory in a file system is the starting point of the entire system, that contains all files and other directories.

E

Envelope   A description of how a characteristic of sound changes over time. Most often, envelopes are used to control the amplitude of oscillators over time, with the most popular amplitude envelope being the ADSR.

F

FFT   An efficient implementation of the DFT.
See: DFT

Float   Data type that represents numbers containing a fractional part, like 3.14.

Frequency Modulation (FM)   Synthesis technique where one oscillator, the modulator, modulates the frequency of another oscillator, the carrier. Used to simulate the vibrato effect, but most notably, used to create a variety of sonic textures with minimal resources (two oscillators).

G

Granular Synthesis   Synthesis technique where a sound source is chopped up in very small segments, which are played back in a cloud-like structure, to either control the pitch or the playback speed, independently of one another, but also to create sonic textures that are completely different from the source.

H

Harmonic   Part of a complex sound, where the oscillation at the base frequency is the first harmonic, and the rest of the harmonics are the overtones. In harmonic sounds, the frequencies of the harmonics are integer multiples of the base frequency. They are also referred to as partials.

Highpass   Filter type that lets the frequencies above its cutoff frequency pass, while it attenuates frequencies below it.

I

IDE   Acronym for Integrated Development Environment. Refers to editors used to write computer code that include features like colour highlighting, word completion, auto-indentation, compilers, and others.

Index   In FM, the index is the amount by which the modulator will modulate the frequency of the carrier. In programming languages, the index is used to locate a value in a list or array.

Integer (Int)   Data type that represents numbers without a fractional part, like 5, 24, or 358.

K

Kwargs   Abbreviation of Keyword Arguments.

L

LFO   Acronym for Low Frequency Oscillator.

Latency   The amount of time it takes for sound to be processed by a computer and output to the speakers.

Lowpass   Filter type that lets the frequencies below its cutoff frequency pass, while it attenuates the frequencies above it.

M

MIDI   Acronym for Musical Instrument Digital Interface. A communication protocol developed in the 1980s, used for communication between hardware.

Modulator   The oscillator that modulates the carrier in various synthesis techinques, like Frequency Modulation, Ring Modulation, Amplitude Modulation, and Phase Modulation.

N

Neural Network A set of artificial neurons that interconnect and exchange numeric data. They consist of the input layer, the hidden layers, and the output layer. In dense feed-forward Neural Networks, each neuron in all layers connects to all neurons in the next layer. Each neuron in all layers but the input includes a weight for each connection, which is multiplied by the input, and the weighted inputs are summed and added to a bias. The result of this is fed to an activation function and then passed to all neurons in the next layer.

Normalize Method that confines the amplitude of a waveform to the audio range, between -1 and 1. Used in band-limited waveforms, and other techniques.

Null A reference that does not currently refer to valid data.

O

OSC Acronym for Open Sound Control. A communication protocol developed in the 1990s, initially as a replacement to MIDI, avoiding the short comings of the latter, eventually used for networked communication between different machines, or between software running on the same machine.

Object An instance of a class. See: Class.

Object-Oriented Programming (OOP) Programming paradigm that employs classes and objects.

Oscillator Tone generators. An algorithm that causes fast changes in values which, when sent to the Digital-to-Analog Converter (DAC), they cause an oscillation in the electric signal sent to the speakers, causing their woofers to move back and forth, thus creating sound.

Oscilloscope A graphic representation of (audio) signals that represent the signal on a 2D plane, where the X axis represents time, and the Y axis represents air pressure (amplitude).

Overtone See: Harmonic.

P

Panning The action of moving the sound in a 2D plane, between speakers.

Partial See: Harmonic.

Phase Modulation (PM) Synthesis technique where one oscillator, the modulator, modulates the phase of another oscillator, the carrier. With PM it is possible to get the same results as with FM.

Phase Vocoder Spectral analysis tool, initially used to alter the duration or the pitch of sound, independently from one another. Now it is used for numerous ways, from cross-synthesis, to spectral filtering or morphing, and others.

Physical computing Physical computing encompasses various Human-Computer Interaction (HCI) practices through the use of sensors that convert analogue signals to digital, transferring information from the physical world to the computer, and vice versa.

Pure Data Open-source multi-media programming environment, following the visual programming paradigm. Developed in the 1990s by Miller Puckette.

Q

Q Abbreviation of Quality Factor. The Q controls the width of the band in bandpass filters.

R

Recursion A method in computer programming where a function can call itself. Useful for compacting code that needs to be repeated. An example of a recursive function is one that outputs Fibonacci series.

Ring Modulation Synthesis technique where two oscillators are multiplied. The perceived sound is two tones with distinct pitches. These pitches are the sum and differences of the pitches of the two oscillators.

S

Sampling rate The rate at which a computer samples amplitude values of audio signals. Typical sampling rates are 44,100Hz (or 44.1kHz), 48kHz, 96kHz, or even 192kHz.

Sawtooth Waveform type where the output signal is a linear ramp that either goes from low to high, or vice versa, and when it reaches its highest point, it resets to its initial value.

Scope Part of code where a variable exists. When a variable is created inside a function, it exists only within the scope of that function, and the rest of the program has no access to it.

Sine wave Waveform type that outputs the sine (or cosine) of a steadily and constantly rising angle, from 0 to 2*pi.

SNR Acronym for signal-to-noise. Refers to the ration between signal and noise, that is determined by the bit depth of an audio system.
See: Bit depth.

Spectrum The range of all audible frequencies, and the energy that is present in each region of it, when a sound is analysed in this respect.

Square wave Waveform type where for half of the cycle it outputs a high voltage (or a 1 in its digital version), and for the other half it outputs a low voltage (or a -1 in digital). If a square wave oscillator includes a duty cycle control, then the latter will control the percentage of one period for which the oscillator will output a high voltage and for which it will output a low voltage.

String Refers to a string of characters in computer code, encapsulated in double or single quotes, depending on the language. An example is: "Hello World!".

Subtractive Synthesis Synthesis technique where a sound rich in harmonic content is filtered, so we can obtain a sound with less harmonics. Typical sound sources for Subtractive Synthesis are sawtooth oscillators and white noise.

SuperCollider Computer music programming language and environment, developed in the 1990s, by James McCarthy.

Synchronous Type of Granular Synthesis, the opposite of asynchronous.
See: Asynchronous.

T

Table-lookup Type of oscillator whose waveform is stored in a table, instead of consisting of an algorithm that creates its waveform by making calculations.

Traceback Stack trace information printed on the console when an error occurs in Python.

Triangle wave Waveform type where the output signal rises linearly for half its cycle, and then falls linearly.

Tuple Python data type that is like a list, but is immutable. A tuple can contain any other data type as its element, including another tuple.

U

Unipolar Refers to oscillators whose amplitude oscillates in one pole, usually the positive one. These oscillators are used for control.

V

Variable A placeholder for a value that varies during the execution of a computer program.

W

Waveform The shape of one period of an oscillator. Typical oscillator waveforms are sine waves, triangle waves, sawtooth waves, and square waves.

Window A shape used to control the amplitude of a sound source in short time segments. Granular Synthesis, or FFT use windowing.

1 Getting Started

Python is one of the most popular programming languages for a number of reasons. One reason is that it is simpler than other languages, like C/C++ or Java. It has a rather simple syntax, it is interpreted – which means that you can run Python line by line, without needing to compile your code before you see the result – and it has a very large community and many different forums, where users can seek help and solutions to their problems. Another reason that makes Python so popular is its incredible number of modules (packages), either native or external, that greatly extend the capabilities of the language. Python is taught at many schools and universities, ranging from primary schools up to Bachelor, Master, or even PhD degrees being achieved using or studying Python.

Python software is actually an interpreter – when you install Python on your computer, you install a program that interprets what you have written in your Python script, and if there are no errors in your code, then this program creates an executable program for every single line in your code. This makes Python slow compared to compiled languages like C or C++ (De Pra and Fontana, 2020), but the benefits of this language usually overshadow this deficiency. For our purposes, we need a language that is very efficient, as audio is a very demanding task with strict temporal requirements. That should make Python not a good candidate for this task, but we will do all our audio processing using the Pyo module. The signal processing parts of this module are written in C and its interface is pure Python (Bélanger, 2016, p. 1214), therefore it utilises the simplicity of Python and the effectiveness of C, bringing together the best of both worlds.

Even though the Pyo module has existed since 2010, its community is rather small and the module remains rather obscure compared to other audio programming environments, like Pure Data or SuperCollider. The existing literature on these environments includes many books that have been written either as educational tools specifically for these languages, like the *SuperCollider Book* (Wilson, Cottle and Collins, 2011) and *Electronic Music and Sound Design* (Cipriani and Giri, 2009), or as tutorials on Digital Signal Processing (DSP) practices using these environments, like *Designing Sound* (Farnell, 2010). Even though Pyo is a very efficient module for DSP, and is part of one of the most popular and versatile programming languages, up to now it lacks a comprehensive guide that covers all of its features and at the same time teaches computer programming and DSP. This book aims to fill this gap by providing an educational tool that targets both beginner and experienced programmers, and both beginner and experienced musicians, from either the electronic or the acoustic music field.

There are many Python modules for audio or DSP, including scipy.signal, librosa, madmom, and sc3. Among these modules, the only one aimed at audio synthesis is sc3, a module to control SuperCollider through Python. The scipy.signal module is aimed more at scientific analysis of signals, whereas librosa and madmom are aimed more at audio analysis than synthesis. This book focuses on audio synthesis with Python, and, concerning the sc3 module, it is advised even

DOI: 10.4324/9781003386964-1

by its developers that users should first learn SuperCollider through its own language, and this, therefore, does not fit our purposes. The Pyo module is entirely based on Python, for both its interface and audio synthesis, without the requirement to learn another language. It is therefore the best fit to learn audio in and through Python.

This first chapter will help the reader install all the necessary software so they can then focus on the programming and creative aspects this book discusses. If you have Python installed on your system, do give this chapter a read, as it covers the installation of the Pyo and the wxPython modules – the latter is used by Pyo for its Graphical User Interface (GUI) widgets. If you have all these modules and you are already programming in Python, you can skip this chapter.

What You Will Learn

After you have read this chapter you will:

- Know how to install and run Python on your system
- Know how to install and use Pip – Python's package manager – on your system
- Know how to search for Python modules in Pip's repositories
- Have installed Pyo and its editor
- Have learned the basic elements you need in an editor
- Have been acquainted with some of the most popular editors

1.1 Installing Python

Python runs on all major Operating Systems (OS), including Windows, macOS, and Linux. On Unix systems like macOS and Linux, Python is installed by default, so there is nothing you need to do. Windows is the only OS that does not include Python by default, though the process of installing it has become much easier in the past few years. Even if Python has not yet been installed on a Windows machine, if you launch the command prompt and type `python`, the Windows Store will launch and you will be able to install it from there.

Ingredients:

- A computer with Windows, macOS, or Linux
- An internet connection
- A terminal emulator (comes with all OSes)

Process:

Python comes in various different versions, with the latest stable one, at the time of writing, being 3.11. There is no real need to use the latest and greatest version of the language, as what we will be doing throughout this book is feasible even with earlier versions. What is necessary though is that you use Python3 and not Python2! Python2 has been declared End-of-Life (EOL) software, and, in 2020, development and support stopped for that version of Python, leaving Python3 the only actively developed and maintained major version.

To check what Python version is installed on our system, we need to open a terminal emulator. On macOS, this is in Applications → Utilities → Terminal.app. On Linux, this is found in the applications menu, or on Ubuntu, you can hit Ctl+Alt+T. On Windows there are a few ways to launch a command line prompt. Either hit Windows+X and click on "Command Prompt", or click "Start" and type "cmd". Once you have a terminal window open, type the following:

```
python --version
```

If you don't get a `command not found` reply, then this will print the version installed. In case this command returns something like `Python 2.x.x`, where x is some number – most likely 2.7.18 or something similar – then the default Python on your system is Python2. If this is the case, or if you do get a `command not found` reply, check if changing `python` with `python3` in the command above gives you Python3. If you don't get a `command not found` reply, then you have Python3 installed, and this is the one you should use.

If the command is not found, then you have to install Python3, either through your system's package manager, or through the official Python website[1]. On macOS and Windows this should be a straight-forward process, as Python's website includes installers for these OSes. As already mentioned, on Windows, typing `python` in the command prompt will launch the Windows Store, and you can simply install Python from there. On macOS you might get a window pop up stating that you have to install the command line developer tools. Just click on the "Install" button, and Python3 will be installed alongside the command line developer tools.

On Linux you will either have to use your system's package manager – e.g. `apt-get install` for Debian and Ubuntu – or download Python's source code and compile it yourself. The source code from the official website comes with the Makefile needed to compile the Python interpreter. All you need to do, as stated in the source's README file, is type the following:

```
./configure
make
make test
sudo make install
```

It is much easier though to just use your system's package manager, as it should be a more streamlined process, and probably the preferred method.

1.2   Installing Python's Package Manager

Python comes with its own package manager that makes finding and installing Python modules a very easy process. This package manager is called Pip. On Linux, Pip is installed by default, so you should not need to do anything. On Windows and macOS, installing Python will also install Pip.

Ingredients:

- Python3
- An internet connection
- A terminal window

Process:

To determine which Python version has Pip installed, type the following line in a terminal window:

```
pip3 --version
```

This will print information concerning the Pip version, but also the Python version for which Pip is installed. Pip is the standard way for installing Python packages – also called modules – even though sometimes we might need to compile a package from its source. Installing a package is as easy as typing the following line:

```
pip3 install numpy
```

When the line above is typed in a terminal, it installs the NumPy module (Harris *et al.*, 2020), a module for scientific computing that is widely used among Python programmers. We might need to install a Python package for a specific Python version instead of the default. This scenario is possible when you have the latest Python as your default, but a module you need has no wheels – is not available through Pip – for this Python version. Installing a module for a specific Python version is possible when Pip is invoked through that Python version. The example line below installs NumPy for Python 3.10:

```
python3.10 -m pip install numpy
```

1.2.1   Searching Modules with Pip

In the past, it was possible to run the `pip search packagename` command, to look up the Pip repositories for a specific package. This is not possible anymore, and there are currently no plans to revive this. To look up modules, we can visit Pip's website[2] and search there. A Python module has also been developed that enables searching through the terminal, the way it was possible to search with the `pip search` command. This module is called pip-search and can be installed with `pip install pip_search`. Once installed, to run it, type the following line in a terminal window:

```
pip_search packagename
```

A list with all the relevant modules will be printed on the console. If this doesn't work, you might need to invoke the module through Python, by typing the following line:

```
python3 -m pip_search packagename.
```

1.2.2   Uninstalling Modules with Pip

If we need to uninstall a Python module that we have installed with Pip, all we need to do is type the following line:

```
pip3 uninstall packagename
```

This command is likely to invoke the removal of more packages than the one specified, as many Python modules do come with dependencies that are installed by the package manager.

It is therefore likely that you will be asked if you are willing to uninstall certain packages that are related to the one you want to uninstall. Usually this should not cause any problems, so you should go ahead and uninstall everything Pip suggests.

1.3   Installing an Integrated Development Environment

To write Python code you will need a text editor, but preferably, an Integrated Development Environment (IDE). IDEs include features like colour highlighting certain keywords of the programming language used, code completion, and ways to execute the programs you are writing.

Ingredients:

• A computer with Windows, macOS, or Linux
• An internet connection

Process:

There are a lot of IDEs designed for Python, or including Python in their features. Perhaps, the most popular Python IDE is IDLE, which is Python's official IDE. IDLE is a simple Python IDE but it does provide features that can facilitate the development process of your projects. Its main advantage is that it has both a shell window and an editor window. In the editor window you can write your Python code, and in the shell window you can run it. Having both, side by side, can speed up the process of development, as you can easily test your scripts frequently, spot bugs, or enhance features as you go. When you install Python from the installers provided by Python's official website or from your system's package manager, IDLE is also installed on your system, so you don't need to do anything else. On Linux, IDLE is not installed by default, but you can easily install it from your system's package manager by typing the following line on Debian, Ubuntu, and other Debian based distributions:

```
apt-get install idle
```

Apart from IDLE though, there are many more IDEs you can use with Python. Actually, you can use any text editor to write your scripts, and then run them from a terminal window. Using a generic text editor though will lack the features an IDE has, like colour highlighting, automatic indentation, code completion, and others, so you are strongly encouraged to use an IDE that has these features. Other popular IDEs used for developing in Python include PyCharm, Visual Studio, and Sublime Text. At the time of writing, one very popular editor, Atom, has been sunset by its developer team. The Pulsar editor is a replacement of Atom that you might want to consider, but at the time of writing, it is still very new, with features still being developed.

One thing to note when it comes to editors and a specific programming language, is the support of this editor for the given language. Python being very popular, is supported by most editors for programming. Another feature that is convenient when coding in Python is launching scripts from the editor. IDLE, Visual Studio, Sublime Text, and PyCharm can all run your Python code. Of course, Vim and Emacs are among the text editors used by Python programmers, but, even though these are very flexible editors, learning how to use them is a rather involved process that is beyond the scope of this book. Therefore, they are mentioned only in the interest of completeness. Regardless of which editor you choose to use, make sure to install any available

Python packages – e.g. a Jupyter Notebook extension can be very helpful when developing in Python.

Another editor that will be useful throughout this book is E-Pyo, Pyo's own editor. This editor is written entirely in Python and includes features targeted at Pyo usage. Compared to the rest of the editors, E-Pyo lacks many features, like word completion, or code documentation pop-up, and it might seem inferior. It does include many audio examples, though, plus you can run your Python/Pyo code from the editor, so it is definitely worth giving it a try. The E-Pyo script comes with Pyo, whether you install the module with Pip or you compile from sources. Since this is just a Python script, there is no need to install anything. All you need to do is run the script by typing the following line in a terminal window:

```
epyo
```

If you use an editor that cannot launch Python scripts, you can still launch scripts from a terminal window, using the `python3` command. If you specify the Python file to be launched, then your Python program will run. If you run this command without any arguments – any extensions to the command – the python interpreter will launch and you can write Python code interactively. All this is explained later on in this chapter.

1.4   Installing Pyo

This book is entirely based on Pyo, so we have to install this module. Pyo is included in Pip's repositories, but it is also open-source, so it is possible to compile it from source. In this section we will look at how we can install Pyo in our system with both approaches.

1.4.1   *Installing Pyo with Pip*

The easiest way to install Pyo is to let your system's package manager install it for you. Compiling from source is not a difficult task, but using the system's package manager is generally the preferred method.

Ingredients:

- Python3
- Pip
- An internet connection
- A terminal window

Process:

Installing a Python module with Pip is extremely easy. Pyo's documentation[3] covers how to install it this way. The only caveat with this approach is that Pyo will not always have wheels for the latest Python version. At the time of writing, Python's latest version is 3.11, and the latest Pyo does support this. It is very likely though that you will find the two not always synced. If you want to install it through Pip, you can either install it with the Pip version that launches with the `pip3` command, or you can target a specific Python version, by typing the following line in a terminal window:

```
python3.10 -m pip install --user pyo
```

This will install Pyo for Python3.10 on your system. Read Pyo's documentation page for more information. As already mentioned, Pyo includes a set of GUI widgets that are written with the wxPython module, including the E-Pyo editor. To be able to use these widgets – we will be using them often throughout the book – you will have to install this module as well. Make sure to use a Python version that has wheels for both modules, Pyo and wxPython. It is likely that this version is not the latest Python.

1.4.2 *Compiling Pyo from Sources*

If you want to install Pyo in a newest Python version that has no wheels for it yet, you will have to compile Pyo from sources. Discussing the process of compiling software in detail is beyond the scope of this chapter, so we will only go through the steps without much explanation. Still, don't get discouraged if you have never compiled software before, it is a very easy process.

Ingredients:

* Python3
* An internet connection
* A terminal window

Process:

Pyo is open-source and the code can be found on GitHub.[4] Pyo's documentation[5] includes instructions on how to compile it from source. There are a few things to consider before attempting to compile Pyo. First of all, if you want to use Pyo's GUI – and you are strongly encouraged to use it – you should make sure you have the wxPython module installed in your system, for the Python version you will compile Pyo against. If you want to compile Pyo for a Python version that has no wheels for wxPython, head over to wxPython's website[6] and get the sources, as you will have to compile that yourself too.

Another thing you must decide is which audio server you want Pyo to compiled against. On macOS you might want to use CoreAudio, and on Linux or macOS, you might want to use Jack. The following line is the command to compile Pyo without specifying an audio server:

```
sudo python3 setup.py install
```

Run this command from the root directory of the Pyo source. Make sure to use the Python version you want to compile Pyo against – Python3.11, or simply Python3, depending on the version. If you want to specify the audio server, you have to include the corresponding flag, so the command above can be modified as in the line below:

```
sudo python setup.py install --use-jack
```

If you want to compile Pyo with 64-bit resolution – note that this refers to the sample resolution, not the resolution of your system – you have to include the `--use-double` flag. If you want to combine the 64-bit flag with the audio server flag, you have to compile with the following line:

```
sudo python3 setup.py install --use-double --use-jack
```

Pyo includes some scripts that take care of all of this for Linux and macOS. Check the documentation page, or the scripts/ directory that comes with Pyo, for further information.

Even though using your system's package manager to install software is often the preferred method, compiling software yourself can provide greater flexibility. At the time of writing, installing Pyo with Pip installs Pyo without Jack support. If you really need to use Pyo with Jack – and it is likely that this is the case when using Linux, for example – you will have to compile it yourself, so you can customise the installation according to your needs.

1.5   Coding Conventions

Throughout this book, the code will be referred to either as a script, or a shell, numbered according to the chapter number. Below we have Script 1.1, the first script of the first chapter. A script is supposed to be written in a text editor and saved to a file with the .py suffix – e.g. myscript.py. This script is pre-defined and can be run either through your IDE's Python command, or from a terminal window, with the `python myscript.py` command.

Script 1.1 A Python script.
```
1  def greet(person):
2      print(f"hello {person}")
```

Below this paragraph we have Shell 1.1, the first shell of the first chapter. Shells refer to an interactive Python session. Such a session can be launched either with the `python` command without any arguments in a terminal, in an IDLE shell, using a Jupyter Notebook either inside your IDE or in your browser, or in some other way that your setup might support – e.g. the Pulsar editor includes the Hydrogen package, a former Atom package, that enables interactive Python coding in the editor. The leading prompt ">>> " in Shell 1.1, is the prompt in a terminal or IDLE session. Depending on how you use Python interactively, this might differ.

Shell 1.1 An interactive Python session.
```
>>> import numpy as np
>>> a = np.arange(10)
```

The difference between a script and an interactive shell session is that with the latter, every typed line is immediately interpreted, enabling interactive programming. A script on the other hand, has to be saved in a file and invoked with the Python interpreter. A script will run from top to bottom and exit, unless we define a way to keep the script alive until we quit it explicitly.

A shell might seem more flexible, as coding is interactive, but what is typed in a shell cannot usually be changed. So, if a mistake in the code occurs, the user can't edit the code, so they will have to re-write it. A script on the other hand is not interactive, but the code is editable. An in-editor Jupyter Notebook or another way to run Python code interactively inside an editor is a feature that can prove to be very helpful, especially when it comes to debugging code, or when trying to get calculations right.

Throughout this book, code will be shown in scripts and shells inter-changeably, though for the most part, it will be shown in scripts. When running an interactive shell in a terminal, on Linux and macOS we can type Ctl+D to exit it. On Windows we can type Ctl+Z and then hit the Return key (the Enter key) to exit, or call the `exit()` command. When running a script that doesn't exit because it has been explicitly programmed to stay alive, or because it gets stuck, we can type Ctl+C to force quit it if it has been launched from a terminal.

There are some cases in this book where a line of code is too long either because of the names of variables or functions, or because of indentation, and it has to be broken to fit these pages. This is not necessarily the case when writing code in an IDE. Throughout this book, such long lines are broken in more, shorter ones. Script 1.2 shows some examples with pseudo-code.

Script 1.2

```
 1  some_module_with_a_long_name.some_long_named_function(
 2      a_long_named_argument,
 3      another_long_argument
 4  )
 5
 6          an_indented_function_name(first_argument,
 7                                   second_argument)
 8
 9          an_indented_variable = some_value *\
10                                 another_value
```

In some chapters we will visit code from other programming languages and environments. As with the Python code, it will be headlined and numbered, like the script and shell above. Depending on the language used, a different terminology will be used instead of script or shell, like sketch for Arduino code. By the time you get to these chapters you will have become accustomed to the way code is illustrated in this book.

1.6   Conclusion

We have installed Python, Pip, Pyo, at least one text editor, and probably wxPython. We have also learned about different Python versions, how to distinguish them in our system, and how to target a specific Python version when installing modules. We have seen how we can install and uninstall a package with Pip, and how to compile Pyo from its source. We are now ready to move on to the next chapter and start writing code!

Notes

1 www.python.org/
2 https://pypi.org/
3 https://belangeo.github.io/pyo/download.html
4 https://github.com/belangeo/pyo
5 https://belangeo.github.io/pyo/compiling.html
6 www.wxpython.org/

Bibliography

Bélanger, O. (2016) 'Pyo, the Python DSP Toolbox', in *Proceedings of the 24th ACM International Conference on Multimedia*. New York, NY, USA: Association for Computing Machinery (MM '16), pp. 1214–1217. Available at: https://doi.org/10.1145/2964284.2973804.

Cipriani, A. and Giri, M. (2009) *Electronic Music and Sound Design*. Rome: ConTempoNet s.a.s.

De Pra, Y. and Fontana, F. (2020) 'Programming Real-Time Sound in Python', *Applied Sciences*, 10, p. 4214. Available at: https://doi.org/10.3390/app10124214.

Farnell, A. (2010) *Designing Sound*. Cambridge, Massachusetts: MIT Press.

Harris, C.R. *et al.* (2020) 'Array programming with NumPy', *Nature*, 585(7825), pp. 357–362. Available at: https://doi.org/10.1038/s41586-020-2649-2.

Wilson, S., Cottle, David and Collins, N. (2011) *The SuperCollider Book*. Cambridge, Massachusetts: MIT Press.

2   Writing Your First Python Programs

In this chapter we will learn how to write small Python programs, to undertake various tasks of varying complexity. By the end of this chapter you will be able to write Python scripts, using code structures and functions to facilitate your development. If you are competent in Python programming, you can skip this chapter and move on to the next, where we will learn various audio synthesis techniques. If you are new to Python, this chapter will get you up and running so you can follow the rest of this book.

What You Will Learn

After you have read this chapter you will be able to:

- Write simple Python scripts
- Identify and choose the appropriate data types for your needs
- Write code structures including conditional tests and loops
- Write your own functions
- Read and understand Python's traceback
- Apply try/except blocks
- Import modules in your scripts
- Create sound with Python
- Use Pyo's Graphical User Interface
- Combine scripts with live input from the interpreter

2.1   Hello World!

Our first recipe will be the famous "Hello World!" program. This is the most common program when starting to learn most programming languages. All this program does is to print "Hello World!" on your computer's console. Still, this very small and simple program can provide information about how code is structured in a given language, how it is compiled, and how it is run. Let us write this program in Python.

Ingredients:

- Python3
- A text editor (preferably an IDE with Python support)
- A terminal window in case your editor does not launch Python scripts

DOI: 10.4324/9781003386964-2

Process:

The easiest way to write the "Hello World!" program in Python is the shown in Script 2.1.

Script 2.1 Simple "Hello World!" Python program.
```
1 print("Hello World!")
```

Write the line in Script 2.1 into a file called "hello_world.py" (or whatever name you want to give it, just make sure the .py suffix is there), and run it. Sure enough, you will see Hello World! printed on your console. Note that in a terminal, you have to be in the directory where you saved your script, so that Python can find it and run it. Depending on the editor you use, there might be other ways to run Python scripts. In the E-Pyo editor you can run a script from the Process→Run menu, or with the Ctrl+R shortcut. This one-liner program is totally legal in Python. Do mind though, that the way we have written it does not convey proper program structuring. To write a better "Hello World!" program in Python we have to include our one-liner into a function, and call it. Functions in Python are explained further on in this chapter, so let's move on.

2.2   Comments in Python

Our next recipe will explain comments. Many languages provide to capability to include comments in your code. Comments are ignored by the compiler and are useful for programmers to either convey information to other programmers about specific parts of their code that might not be very easy to understand, or to keep some notes for themselves. When returning to a program you wrote a while back, you might find yourself having difficulty remembering, or even understanding what certain parts of your code do. Inserting comments in certain parts of your program will make this process much easier.

Ingredients:

- Python3
- A text editor (preferably an IDE with Python support)
- A terminal window in case your editor does not launch Python scripts

Process:

In Python comments start with # and have no effect on the script. For example, the line in Script 2.2 is our print() program, including a comment that is completely ignored by the Python interpreter.

Script 2.2 A comment in Python.
```
1 print("Hello World!") # prints "Hello World!" on the console
```

If you run this script, you will get exactly the same result as with our first recipe. This comment explains what this line of code does. This specific example is trivial, but is enough to explain what comments are, and how and why they are used.

2.3 Data Types in Python

Python includes different data types that can be assigned to variables. These can be roughly categorised in the following groups: numeric, Boolean, sequence, dictionary, and set.

2.3.1 Numeric Data Types

The first data type we will look at is the various numeric data types. This data type group consists of three members, integer, float, and complex. The integer data type is used for storing integer values, like 1, 2, 3, -5, 1000000. To improve readability, we can separate groups of digits with underscores. For example, 1000000 can be written like this: 1_000_000. These two numbers are exactly the same, one million. In programming jargon, integers are called ints.

The other numeric data type is float. This data type stores non-integer values like 3.14, 100.5897, and -5.893459. Python includes a third numeric data type, called complex. This data type stores complex values that include a real and an imaginary part. Such a value can be expressed like this: 2 + 3j, where the second value is the imaginary. We will not be using complex numbers in this book at all.

Ingredients:

- Python3
- An interactive Python shell

Process:

Numeric values are usually stored in variables which, as a good programming practice, are given self-explanatory names. For example, if we want to store the value of a hypotenuse, we could name the variable that will hold this value `hypot`, or something similar. The pseudo-code in Shell 2.1 is an example of self-explanatory variables holding some float values.

Shell 2.1 Self-explanatory variable names.
```
>>> side_a = 3.0
>>> side_b = 2.0
>>> hypot = square_root(side_a*side_a + side_b*side_b)
```

While this is pseudo-code, it is almost valid Python code. If you run this code, though, you will get an error message when you execute the third line. This specific example is a bit more involved, and it is used here as an example only, to explain the role of variables and how to make them self-explanatory.

In many programming languages, mixing data types is only possible through what is called casting. In Python it is possible to combine different numeric types without explicitly changing the types of variables. The code in Shell 2.2, being perfectly valid, stores some arbitrary values of all three numeric data types, and combines them. Launch a Python shell and write the lines in Shell 2.2 without the comments, hitting the Return key at the end.

The first three variables have somewhat explanatory names, as we named the integer `i`, the float `f`, and the complex `c`. The rest of the variables take arbitrary names for the sake of simplicity, since this example only explains the numeric data types of Python, and these values have no actual functionality.

There two things to note here. The complex data type has two values, both of which are floats, even though we define these values as integers. We can access them separately using the `real` and `imag` members of this data type. The second thing to note is that if we add a float to an integer, and store the result to the latter, it switches from integer to float, as in the last line of Shell 2.2.

Shell 2.2 Python numeric data types.
```
>>> i = 5 # this is an int
>>> f = 1.5 # this is a float
>>> c = 3+5j # this is a complex
>>> a = i + f # this is 6.5 and it is a float
>>> b = f + c.real # this yields 4.5, 1.5 + the real part of c
>>> d = i + c.imag # this yields 10.0, a float
>>> i = i + f # now i is 6.5, a float
```

2.3.2   Boolean Data Type

The next data type we will look at is the Boolean. This is a standard data type in many programming languages. It is a true/false value, essentially a 1 and a 0. Booleans are used in conditional tests to determine whether a part of our code will be executed or not. Python reserves the keywords `True` and `False` for the Boolean data type.

Ingredients:

- Python3
- An interactive Python shell

Process:

We can create a Boolean variable by simply assigning one of these two keywords to it. Shell 2.3 is an example. Note that arithmetic operations with Booleans are valid in Python. The lines in Shell 2.4 are perfectly legal. When we make arithmetic operations with Booleans, the latter are treated as integers with the values of 1 for `True` and 0 for `False`.

Shell 2.3 Python's Boolean data type.
```
>>> a = True
```

Shell 2.4 Arithmetic operations of Booleans in Python.
```
>>> a = True
>>> b = a + 1 # this results in 2 and b becomes an int
>>> c = a * 2 # this results in 2 as well
>>> d = a / 2 # this results in 0.5 and d becomes a float
```

2.3.3   Sequence Data Types

The third data type we will look at is the sequence data type. This group includes strings, lists, and tuples.

Ingredients:

*   Python3
*   An interactive Python shell

Process:

The string data type is common among many programming languages, with Python being among them. We will look at this data type first.

2.3.3.1   The String Data Type

In computer programming, a string is a sequence of characters. Shell 2.5 shows a Python string. If you type it in a Python interpreter, you will see it being echoed back on the console. Usually we store strings in variables, in which case there is no echoing, or we pass them to functions as arguments (more on functions and arguments later on).

Shell 2.5 A string in Python.
```
>>> "this is a string"
```

In Python, strings are encapsulated either in double or single quotes. If we want to include quotes inside our string, so these are printed, we have to use the quote sign that is not used to encapsulate our string. For example, the string in Shell 2.6 is valid. This string can also be written with the quotes inverted, meaning that we can encapsulate our string in single quotes, and use double quotes around a.

Shell 2.6 A string containing quotes.
```
>>> "the value of 'a' equals 5"
```

Strings are typically used for printing information on the console, or to convey arbitrary information that can be expressed more easily with a natural language, instead of numeric data types. Since Python3.6 there is a special type of string, call the f-string. This string type allows us to insert expressions inside strings. For example, the code in Shell 2.7 including an f-string is valid.

Shell 2.7 A Python f-string.
```
>>> a = 5
>>> print(f"a+1 equals {a+1}")
```

Shell 2.7 will print "a+1 equals 6" as the {a+1} will print the value of the expression of the a variable + 1, which is 6. We will be using this kind of string formatting throughout this book, so you are strongly encouraged to use Python3.6 or greater.

2.3.3.2   The List Data Type

A list is a sequence of arbitrary data types. A single list can contain numeric values (all three types), strings, even other lists. In Python, lists are denoted with square brackets. The examples in Shell 2.8 are all valid Python lists.

Shell 2.8 Various lists.
```
>>> lis1 = [1, 4, 19.7, 5_900]
>>> list2 = ["a", "list", "of", "strings"]
>>> list3 = [54, list1, 178.4678, list2]
```

Notice that we included the first two lists in the third one. This is perfectly valid and, in many cases, very effective. A list can also be updated with new elements. The code in Shell 2.9 defines an empty list, and then inserts a few elements in it. The resulting list will have the following elements: $0, 5.48$, "foo", $[1,2,3]$.

Shell 2.9 Appending items to a list.
```
>>> l = []
>>> l.append(0)
>>> l.append(5.48)
>>> l.append("foo")
>>> l.append([1,2,3])
```

2.3.3.3 The Tuple Data Type

The last sequence data type is the tuple. This is very similar to the list, but it is immutable. A tuple can contain any kind of data types, like the list does, but once it is defined, its elements cannot change, neither can the tuple be updated, like the list can. Tuples are denoted with round brackets. Shell 2.10 includes some examples of valid tuples:

Shell 2.10 Tuples.
```
>>> t1 = (1, 2, 5.6)
>>> t2 = ((6, 98), "foo", [5, 9, "bar"])
```

2.3.3.4 Indexing Sequence Data Types

As in many other programming languages, in Python indexing sequence data types is zero-based. This means that the first item of such a data type has the index 0, the second item has the index 1, and so on. Indexing sequence data types is done with square brackets. Shell 2.11 includes some examples of indexing.

Shell 2.11 Indexing in Python.
```
>>> l = [100, 54, 9, [8, 7]]
>>> print(l[0]) # this will print 100
>>> print(l[1]) # this will print 54
>>> print(l[2]) # this will print 9
>>> print(l[3][1]) # this will print 7
>>> s = "this is a string"
>>> print(s[3]) # this will print "s" from the word "this"
>>> t = (4, 78, "foo")
>>> print(t[2][0]) # this will print "f"
```

As we can see, indexing can go as deep as the items we want to index. The fifth line accesses the fourth item of the list l, indexed with 3, which is a list with two values, 8 and 7, and then accesses the second item of this sub-list, indexed with 1, the value 7. The last example accesses the third item of t, which is indexed with 2 and contains the string "foo", and the first character of this string, indexed with 0, which is the letter "f".

2.3.4   The Dictionary Data Type

In Python, it is possible to group values together and index them in a more arbitrary way instead of zero-based sequential indexes, like the ones used with lists, strings, and tuples.

Ingredients:

- Python3
- An interactive Python shell

Process:

Python includes a special data type called a dictionary. Dictionaries are a bit like lists, only in this case, we define the index together with the value the index is pointing to. In this case, indexes are called keys. Dictionaries in Python are denoted with curly brackets. Their keys are separated from their values with a colon, and key and value pairs are separated with a comma. The example in Shell 2.12 is a valid Python dictionary.

Shell 2.12 A dictionary.
```
>>> d = {"one": 56, "two": [0, 6, 2], 5: "foo"}
```

The keys we have defined in this dictionary are "one", "two", and 5. Note that keys consist of both strings and ints. Data types that can be used for dictionary keys include ints, strings, floats, and Booleans, though, numeric data types might not be so meaningful in dictionaries, especially ints and Booleans, as ints are used to index lists and tuples, and Booleans are more or less treated likes ints. On the other hand, dictionary keys are not incremented like they are in the example above of Shell 2.11. In Shell 2.12, the third key is 5, whereas in a list or tuple, the third item would be indexed with 2.

We can thus see that dictionary keys can be created arbitrarily, to fit our needs. Nevertheless, dictionary keys are most often strings. The values of dictionaries are accessed through their keys. The dictionary in Shell 2.12 can be accessed the way it is shown in Shell 2.13.

Shell 2.13 Accessing dictionary values.
```
>>> d["one"] # this yields 56
>>> d["two"][1] # this yields 6
>>> d[5] # this yields "foo"
```

Dictionary values can be changed, or new key/value pairs can be inserted. The lines in Shell 2.14 are valid. The first line changes the value of the key 'one', and from the integer 56, the value is now the string 'bar'. The second line adds a new key/value pair to our dictionary, with the key 'three' and the value 89.

The dictionary data type is very convenient in many cases, and we will encounter them a few times throughout this book.

Shell 2.14 Changing and adding dictionary values.
```
>>> d["one"] = "bar"
>>> d["three"] = 89
```

2.3.5   *The Set Data Type*

The last data type we will look at is the set. This is an unordered collection of data types that is iterable, mutable, and has no duplicated elements. We will sparsely use sets in this book, so we will explain them here in brief.

Ingredients:

- Python3
- An interactive Python shell

Process:

A set is denoted with curly brackets, a bit like the dictionary, but it only includes values and no keys. A set is useful though if created from a string or a list, so duplicated elements can be easily discarded. Shell 2.15 shows some examples of sets.

Shell 2.15 The set() datatype.
```
>>> myset = {"a", 4, 8, "foo"} # manually created set
>>> l = [5, 3, 7, 5]
>>> newset = set(l) # this results in {3, 5, 7}
>>> s = "this is a string"
>>> lastset = set(s)
>>> # the set above is { 'g','n','s','i','a',' ','h','r','t'}
```

Note that the order of the elements of the list or the string we used to create a set was not maintained. In the case of the string, each character was separated and used only once, so we get only one "s" character.

Accessing items of sets is done either with loops, or by converting a set to a list, and then using normal list indexing. Shell 2.16 demonstrates this. Note that we will talk about loops further on in this chapter, so don't worry if you don't really understand how the loop in this shell works.

Shell 2.16 Accessing set items.
```
>>> for item in myset:
...     print(item)
...
>>> list(newset)[1] # this yields 5
```

This loop results is printing all the elements of myset sequentially. Note that after hitting the Return key after typing the colon, the prompt will change from three > characters, to three dots. This is a prompt of code structures in the Python interpreter. These prompts are not present in text editors, but only when we write Python code inside the interpreter in a terminal or in IDLE. Also note that after the print(item) line, you must hit Return twice, to exit the loop body. The last line accesses the second element of newset. Sets include operations to remove or insert items. See the example in Shell 2.17.

Shell 2.17 Adding and removing set items.
```
>>> myset.remove('foo')
>>> newset.pop() # this removes the first item, in this case 3
>>> lastset.add('l') # adds 'l' to lastset
>>> lastset.add('s') # has no effect as 's' is already included
```

2.3.6 Reserved Data Type Keywords

Each data type in Python has a reserved keyword that can be used to create such a data type. The following keywords are reserved by Python, and can be used to initialise a variable with the given data type:

```
int(), float(), complex(), str(), list(), dict(), tuple(), set()
```

Note the round brackets used with the keywords. This is because all these types are actually classes in Python, and the variables of these classes are called objects. We will learn more about classes and objects in Chapter 8, but bear in mind that these two terms will be used interchangeably throughout this book.

Ingredients:

- Python3
- An interactive Python shell

Process:
We can initialise a variable of any of these data types by using the respective keyword, as shown in Shell 2.18. The first two lines initialise variables with the value of 0 assigned to them, while the third line initialises an empty dictionary and the fourth an empty list. The last line initialises an empty tuple. This is probably redundant, as tuples are immutable, so once initialised, no modifications to it are possible, so we will be left with an empty tuple.

Shell 2.18 Initiating empty data types.
```
>>> a = int()
>>> b = float()
>>> d = dict()
>>> l = list()
>>> t = tuple()
```

2.3.7 NoneType

Python has a special data type that is equivalent to the Null data type of other languages. This type is called a `NoneType` and inside our code we access it by typing `None`. This is a neutral data type that holds no actual value.

Ingredients:

- Python3
- A text editor (preferably an IDE with Python support)
- A terminal window in case your editor does not launch Python scripts

Process:

The `NoneType` can be useful for initialising variables whose data type is still unknown, or when we want to check if an operation succeeded or failed. In case it fails, such an operation could return a `NoneType`, so we could test against it to see what happened. Script 2.3 is a pseudo-code example.

Script 2.3 The None data type.
```
1 a = some_operation()
2
3 if a is None:
4     print("operation failed")
5 else:
6     print("operation succeeded!")
```

This is pseudo-code, as the name of the `some_operation()` function, which supposedly returns None in case of failure, is made up. We will see a more concrete example of a function that can return None later on in this chapter. This code also includes a conditional `if/else` test, which we will see next, plus the `is` reversed keyword, which tests if two variables are of the same data type, or if a variable is of a certain data type, like it is used in the code above.

2.3.8 *Data Types Are Classes*

Even though we referred to instances of data types as variables, since their content varies, these elements are actually what in Python and other languages is called an object. All the data types we learnt in this section are different classes. The code in Shell 2.19 demonstrates that.

Shell 2.19 Data types are classes.
```
>>> i = 5
>>> type(i)
<class 'int'>
>>> d = dict()
>>> type(d)
<class 'dict'>
```

We can see that the result of `type(i)` prints `<class 'int'>` telling us that the variable i is an instance of the class `'int'`. Python is an Object-Oriented Programming language (OOP), and in programming jargon, the instance of a class is called an object of that class. We will see objects in many places throughout this book, even in this chapter, so it is good to get acquainted with this terminology from an early stage. Do keep in mind though that when creating objects of the data types we covered in this section, we will still refer to them as variables and not objects. Syntax conventions in Python will help you differentiate what is called a variable and what an object, as usually we refer to a class instance as an object when the class is written with its initial letter in upper case. For now, keep this information in mind and it will start making sense as you read through this chapter and the book.

2.4   Conditional Tests

A standard feature in many programming languages is the conditional test. This is based on Boolean logic, where a condition is either true or false. In Python, like many other languages, this is done with the `if` structure.

Ingredients:

- Python3
- A text editor (preferably an IDE with Python support)
- A terminal window in case your editor does not launch Python scripts

Process:

If we want to print something on the console if a value is above a threshold, we can do it the way it is shown in Script 2.4.

Script 2.4 An if test.
```
1 val = 5
2 if val < 10:
3     print("val is less than 10")
```

If we save this code to a script and then run it, we will have the string passed to `print()` printed on the console. Of course this specific conditional test is rather useless, as the value of the variable `val` is hard-coded and we know the result of the test in advance. This is only a demonstration to show how the `if` test works in Python. This test is used with variables that change while a program runs, or with variables that are not known prior to this test (e.g. a variable that gets its value from input from the user).

What is important is to learn the structure of this test, where the `if` reserved keyword is followed by the condition (in this case, `val < 10`) and then a colon. After the colon we write the body of the code to be run in case the condition is true. If this body of code is a single line, it can be written in the same line with the condition, right after the colon. Otherwise it has to be written one line (or more) below.

What is important is the indentation. Indentation is very important in Python, as this language does not use curly brackets or some other way to encapsulate code snippets. The amount of indentation is up to the user, but it should be consistent throughout the entire script. This can be done either with horizontal tabs or sequential white spaces. By convention, indentation in Python is one horizontal tab. Many text editors that support coding in Python import these tabs automatically.

The code in Script 2.4 runs only when our condition is true, but we can have more code that will be executed if our condition is false. The code in Script 2.4 can be extended to the code in Script 2.5.

Script 2.5 An if/else test.
```
1 val = 5
2 if val > 10:
3     print("val is greater than 10")
4 else:
5     print("val is less than 10")
```

Now we have changed the code a bit and we are testing if `val` is greater than 10, instead of less. This condition is false, so our test will fail. In this case, the code of `else` will be executed. We can extend our code though even more and include more cases, in between `if` and `else`. This is done with the `elif` reserved keyword, which stands for "else if". So our code now changes to the code in Script 2.6.

Script 2.6 An if/elif/else test.

```
1 val = 5
2 if val > 10:
3    print("val is greater than 10")
4 elif val == 5:
5    print("val is equal to 5")
6 else:
7    print("val is less than 10 and not equal to 5")
```

This script will print "val is equal to 5" on the console, since `val` is equal to 5, and this is the test of our `elif`. In a conditional test, `if` and `else` can be used only once, with the former at the beginning and the latter at the end. `elif`, though, can be used as many times as you want, and these tests should be in between `if` and `else`.

Note the double equal sign used in the `elif` test. This is a test for equality and should not be confused with the single equal sign which is used for value assignment. The two lines in Shell 2.20 are completely different, but sometimes easy to be mixed.

Shell 2.20 The assignment and test for equality signs.

```
>>> val = 5
>>> val == 5
```

The first line assigns the value of 5 to a variable called `val`, and the second line tests if `val` is equal to 5. The first line, apart from assigning a value to the variable `val`, defines this variable, in case it has not yet been defined. Note that if no value assignment has been made to `val`, then using the second line will produce an error, since the variable `val` will have not yet been defined.

2.5   Functions in Python

A feature that is common among many programming languages is a function. Python is no exception to this rule, and we will be using functions throughout this book.

Ingredients:

- Python3
- A text editor (preferably an IDE with Python support)
- A terminal window in case your editor does not launch Python scripts
- An interactive Python shell

Process:

A function is a set of computations that might take some input or not, and they produce some output. For example, a simple function that adds together two numbers should take two input values and produce one output. Such a function is defined in Shell 2.21.

Shell 2.21 A two operand addition function.

```
>>> def add(a, b):
...        return a + b
...
>>>
```

The keyword `def` is a Python reserved keyword that stands for "definition", and it implies a function definition. After this keyword, we use a word that we wish to use as the name of the function we are defining. In this case, we call our function `add`. After the name of the function we have to place a pair of round brackets. In case our function takes input, we include that input here. In programming jargon, this input is called arguments. Since we are adding two numbers, we have to insert two arguments, and when defining our function, we express these arguments with arbitrary variable names. In our simple case, we call our variables `a` and `b`. After the round brackets we have to type a colon and then we go one line below and we write the body of our function. Note that the second line is indented, the same way code is indented after an `if` test.

The body of our function is one line only. This line returns the addition of the two arguments. This is achieved by using the reserved keyword `return` and then the expression we want our function to return, the addition of the two arguments. As with the loop example in the `set()` data type section of this chapter, since we are writing straight into the Python interpreter, to conclude our function when writing in a terminal window, we must hit Return twice, hence the extra line with the three dots. To call our function, all we need to do is write the line in Shell 2.22.

Shell 2.22 Calling the addition function.
```
>>> add(3, 5)
```

Since we are writing this in an interactive shell, the output of the function will be printed on the console, even though we are not using `print()`. This is because our function is returning a value that we are not storing anywhere. When defining functions that return something, it is meaningful to store the returned value in a variable for later use. This is done with the line in Shell 2.23.

Shell 2.23 Storing the output of the addition function.
```
>>> a = add(3, 5)
```

This line calls our function and stores the returned value in a variable called `a`. Note that the first argument of our function is also named `a`, but there is no conflict in this case as the argument `a` of our function is local to the scope of the function, whereas the variable `a` that stores the returned value is local outside the scope of the function. We will talk more about scope further on in this chapter.

To go back to our "Hello World!" program, we can structure it better by including our one-liner in a function that will be called when we run our script. The code in Script 2.7 does exactly that.

Script 2.7 Creating and calling a main() function in Python.
```
1 def main():
2     print("Hello World!")
3
4 if __name__ == "__main__":
5     main()
```

Now we have included a number of standard coding elements, like functions, conditional tests, and a standard way of running Python scripts. In this script we define a function called `main()` which takes no arguments, hence the round brackets are empty. All this function does is print "Hello World!" on the console. Notice that it doesn't return anything, as we don't need to store any value from this function. You might have also noticed that `print()` has a function structure, as it is followed by round brackets with something in them – in this case, the "Hello World!" string. This is because `print()` is actually a function included in Python by default.

Since we put the `print()` function inside another function, we must call the top level function, `main()`, at some place in our program. This happens at the last two lines of our script. These two lines are very standard in Python and they will be explained in detail later on in this chapter.

2.5.1   *Positional and Keyword Arguments*

The arguments we have passed to the function that adds them are called positional arguments. This is because the order in which we write them when we call such a function matters. Python includes another type of argument, called keyword, abbreviated with kwargs (pronounced "quargs").

Ingredients:

- Python3
- An interactive Python shell

Process:

Kwargs are identified by their keyword and we provide them in any order, as their position doesn't matter when calling such a function. Shell 2.24 includes an example of a function with kwargs. They are very flexible as they can be totally omitted, or we can provide only some of them when calling a function with kwargs. Also note one line before the last, where we first provide `arg2` and then `arg1`, still we get `arg1` printed first. In the last line we provide values as positional arguments. This is perfectly valid, and the arguments will be assigned to the kwargs in the order we provide them.

Shell 2.24 A function with kwargs.
```
>>> def func(arg1=1, arg2=2):
...        print(arg1, arg2)
...
>>> func() # prints "1 2"
>>> func(arg1=5) # prints "5 2"
>>> func(arg2=7) # prints "1 7"
>>> func(arg2=59, arg1=9) # prints "9 59"
>>> func(3, 4) # prints "3 4"
```

2.5.2   *Recursive Functions*

In Python, a function can call itself. This practice is called recursion. The code in Script 2.8 is a usual example of how recursion can help calculate a Fibonacci series.

Script 2.8 A recursive function returning the Fibonacci series.
```
1 def fibonacci(n):
2     if n in [0, 1]:
3         return n
```

```
4        else:
5                return fibonacci(n-1) + fibonacci(n-2)
```

Note line 5 where the function is calling back itself, passing its own argument subtracted by 1 and by 2. Take a minute to think about what is happening here. Line 2 tests if the argument is found in the list [0, 1], and if it is, the argument is returned intact, as the first two values of a Fibonacci series are 0 and 1. If the argument is not in this list, then line 5 will be invoked. Let's say that the argument is 2, then line 5 will result in calling the fibonacci(n) function twice, once with the argument 1 and once with the argument 0. These two function calls will result in the values 1 and 0 being returned, as they are in the list [0, 1], and they will be added and returned, so we will get 1 returned. If the argument passed is 3, then this process will be repeated until all recursive function calls are caught by the if test in line 2. So, the function will keep on calling itself as long as the test in line 2 fails.

Recursion in functions can be helpful and can make your code look concise and elegant, but they should be used carefully, as you might get a RecursionError from Python, if you haven't made sure that there is some limit in the function calling back itself. We will learn more about errors in Python in Section 2.8 of this chapter.

2.6   Loops in Python

Python has structures that make parts of our code run in a loop, for a specified number of times. In this section we will look at two different ways to run loops in Python.

Ingredients:

- Python3
- An interactive Python shell

Process:

There are two different ways to run code in a loop, using the for or the while reserved keywords. Let us include the if test we already wrote, in a for loop. The code is shown in Shell 2.25.

Shell 2.25 A for loop.
```
>>> for i in range(10):
...        if i == 5:
...                continue
...        print(i)
...
>>>
```

In this script, we have used the for keyword, combined with range(). range(), which is a class rather than a function (more on classes in Chapter 8), takes from one to three arguments. Here we have passed one argument which is the "stop" argument. This means that range() will iterate over a range from value 0, until it reaches [stop − 1], and it will increment by 1 every time. Essentially, when using range() with one argument only, it will iterate as many times as the argument we pass to it, in our case, it will iterate ten times.

range() returns a list with the values it iterates over. In this code it will return the following list:

```
[0, 1, 2, 3, 4, 5, 6, 7, 8, 9]
```

The for loop will iterate over this list and will assign each of its values to the variable i, every time assigning the next item of the list. Inside the body of our loop we can use this variable for any task we want. This example is very simple and all it does is print on the console the value of i except from the iteration where i is equal to 5. This is achieved by using the reversed keyword continue. This word skips any code of the structure that includes it, that is written below it. So, when our if test is true, the code written below this test will be skipped, and the for loop will continue to the next iteration. Running this script will print the following on the console.

```
0
1
2
3
4
6
7
8
9
```

If the if test with the continue keyword was written after print(i), then it would have no effect at all, since there would be no more code left in the loop to skip.

The same loop can be achieved with the while loop, although we should understand its difference to the for loop. The latter works with a specific number of iterations, while the former works with a conditional test. While the test is true, the loop will run, and when the test is false, the loop exits. We have to be careful when using while as we can create an infinite loop where the conditional test can never be false. The code in Shell 2.26 is an example of an infinite loop.

Shell 2.26 A while loop that never exits.
```
>>> i = 0
>>> while i < 10:
...     print(i)
...
>>>
```

while tests its condition – in this case i < 10 – and if it is true, it will execute the code in its body. After this code is run, while will return back to its conditional test to determine whether it will execute its code again. If the test is false, the loop will exit, and the script will go on to whatever is written after the body of the loop. Note that in the example above, i is set to 0 and while is testing if this variable is less than 10. Since we do not increment i inside our loop, it will always be less than 10, and our while loop will not be able to exit, stalling our program. If you happen to run this loop you can exit it by hitting Ctl+C, as Ctl+D will not work. The for loop we wrote in Shell 2.25 can be written with while the way it is shown in Shell 2.27.

Shell 2.27 The for loop of Script 2.25 implemented with while.
```
>>> i = 0
>>> while i < 10:
```

```
...     if i == 5:
...         continue
...     print(i)
...     i += 1
...
>>>
```

In this code we initialise i to 0 and then we enter the while loop. The conditional test of the loop will succeed since 0 is less than 10 and the body of the loop will be executed. The code of the loop is almost identical to the for loop we wrote earlier, only this time, before we exit the body of our loop, we increment i by 1. The syntax used in the last line of this example is equivalent to the line in Shell 2.28.

Shell 2.28 Expansion of the += semantic.
```
>>> i = i + 1
```

By incrementing the variable by one, we make sure that we will print incrementing values, omitting 5 as we use continue when i equals 5. But most importantly, we make sure that our loop will exit after ten iterations. Omitting to increment the i variable will not only print a constant 0, but will disable our loop from exiting, like with the previous example.

Another functionality of the for loop is to initialise lists, the way it is shown in Shell 2.29.

Shell 2.29 Creating a list with list comprehension.
```
>>> l = [i for i in range(10)]
```

In Python, the square brackets denote a list. The notation above uses the for loop combined with range() to iterate ten times and since it is encapsulated in square brackets, it will return the following list.

```
[0, 1, 2, 3, 4, 5, 6, 7, 8, 9]
```

Initialising lists like this is called list comprehension in Python jargon, and it is a very convenient way to initialise a list with items. We will be using this technique in a few examples in this book.

2.7 Scope

Like in many other languages, Python implements scope, where variables are local to a certain scope.

Ingredients:

- Python3
- An interactive Python shell

Process:

Now that we have learned the basics about functions and loops, we can understand scope. Take the example of Shell 2.30, where we want to create a function that will take an integer as an argument, and it will return this integer added to a constant value.

Shell 2.30 Trying to access a local variable outside of its scope.

```
>>> def func(arg):
...     a = 5
...     return a + arg
...
>>> b = func(a)
```

In this example, the fifth line will throw an error, as we are trying to access the variable a which is defined inside our function and it is local to the scope of that function only. The rest of our code is not aware of the existence of this variable and doesn't have access to it. This variable is created on-the-fly whenever we call this function, it remains accessible only within the function, and it is destroyed when the function exits. Shell 2.31 does not produce any errors as the variable a is in the global scope.

Shell 2.31 Accessing a global variable.

```
>>> a = 5
>>> def func(arg):
...     return a + arg
...
>>> b = func(a)
```

A way to define global variables within a function is by using the reserved keyword global. The code in Shell 2.32 creates a inside the function, but now its scope is global, so we can access it outside of the function.

Shell 2.32 Initialising a global variable inside a function.

```
>>> def func(arg):
...     global a
...     a = 5
...     return a + arg
...
>>> b = func(a)
>>> b += a
```

The way we define the variable a above enables us to both call the function and use the variable outside the scope of this function. Global variables are usually considered a bad practice by many programmers, as being able to access a variable from anywhere inside a program might lead to accidental modifications of it, plus access to global variables is slower than trying to access a local one. Pyo classes, though, run on a different thread than the main thread of Python, and they are often initialised in the global scope, so we will be using global variables in various parts in this book, especially for Pyo classes.

2.8   Traceback

When there is a bug in our code, we need a way to tell where this bug resides. The Python interpreter prints errors on the console when something is not correct. This is called the traceback, and it gives useful information as to which part of our code produced the error, and what type of error that was.

Ingredients:

- Python3
- An interactive Python shell

Process:

Take the code in Shell 2.33 as an example. It will throw an error since we haven't defined the key "age" in our dictionary. The printed traceback is shown in Shell 2.34. The traceback tells us that the error is produced in line 1, and the error type is KeyError, after which the undefined key 'age' is also printed, which is the key that produced the error. The error was spotted in line 1 because we run this example in the Python interpreter, and not by saving and running a pre-defined script. If we write this code in a script and run it, then the traceback would report line 4, and instead of "<stdin>" as the file name, it would print the entire path to the actual Python script we run.

Shell 2.33 Trying to access a dictionary key that does not exist.
```
>>> d = {"name": "Alice", "place": "Wonderland"}
>>> print(d["name"])
>>> print(d["place"])
>>> print(d["age"])
```

Shell 2.34 KeyError traceback.
```
Traceback (most recent call last):
  File "<stdin>", line 1, in <module>
KeyError: 'age'
```

Python has various error types. The most common probably are: IndexError, KeyError, NameError, IndentationError, and perhaps RuntimeError. The names of these errors give us information as to what they relate to. For example, the IndexError is raised when we try to index an item of a list that is out of the bounds of the list. The NameError concerns undefined variables, functions, and classes. The IndentationError includes errors concerning bad indentation. The RuntimeError is special, as this is raised whenever the error occurred does not fall under any of the other error types.

2.8.1 Exceptions

We might find ourselves in a situation where an error is produced at a certain part of our code, causing it to crash, and the traceback doesn't provide enough information to debug it. In Python we can avoid terminating our program when an error is raised, so we can diagnose our code. This is achieved with the exception feature.

Ingredients:

- Python3
- A text editor (preferably an IDE with Python support)
- A terminal window in case your editor does not launch Python scripts

Process:

When a certain type of error is raised, we can modify our code in a way that it will try to execute the code that produces the error, and if an error is indeed raised, we can run an exception. The code in Script 2.9 gives an example.

Script 2.9 Excepting an IndexError.
```
1 l = [1, 2, 3, 4]
2 for i in range(5):
3     try:
4         print(l[i])
5     except IndexError:
6         print(f"i is {i} while l has length {len(l)}")
```

We have defined a list with four values and then we run a loop five times. This would normally cause our program to crash, since trying to access a fifth value would point to invalid memory. What we do though, is try to access the items of l by indexing with the variable i, and if an IndexError is raised, we print some information concerning the value of i and the length of the list l.

Exceptions are very handy when we are debugging our code, but they should be used carefully as abuse of exceptions can lead to code that doesn't work but is almost impossible to debug. One example of our previous code with bad use of exceptions is shown in Script 2.10.

Script 2.10 Exception of all error types.
```
1 l = [1, 2, 3, 4]
2 for i in range(5):
3     try:
4         print(l[i])
5     except:
6         print(i)
```

Here we don't specify the error type we want to except, and we don't print very useful information as to why an error was raised and excepted. In general, exceptions should be used explicitly, specifying the error type we want to except, and never implicitly without a specific error type. Additionally, it is a good idea to provide as much and useful information as possible in an exception block of code, as arbitrary information will not help much in debugging your code.

2.9   Importing Modules and Scripts Inside Other Scripts

Sometimes we have to import native or external modules in our Python script, or we might need to combine more than one scripts into a project.

Ingredients:

- Python3
- A text editor (preferably an IDE with Python support)
- A terminal window in case your editor does not launch Python scripts

Process:

Python comes with a multitude of native modules that are not loaded by default, to keep the language light-weight. Even very standard modules like the random module, have to be imported manually. There are a few different ways to import modules. Script 2.11 demonstrates them.

Script 2.11 Importing modules in Python.

```
1 import random
2 from time import sleep
3 import numpy as np
4
5 a = random.random() # random float between 0 and 1
6 b = random.randrange(5, 10) # random int between 5 and 10
7
8 sleep(0.5) # stall script for half a second
9
10 c = np.arange(10) # numpy array with 10 elements, from 0 to 9
```

In this example we have imported three modules, random, the sleep function from the time module, and NumPy. The random module was imported entirely, and that is because we used a couple of methods from it, namely `random()` and `randrange()`. The former returns a random float between 0 and 1, and the latter returns a random integer within the range given in its arguments.

From the time module we only needed one function, sleep, so we imported only that. This way, instead of having to write `time.sleep(0.5)` we can just write `sleep(0.5)`. The last module we imported is NumPy, a very useful module for numeric calculations in Python. This module is very fast and efficient, and you will probably see it often if you use Python. Since, when using this module we use a lot of its methods in many places in our code, we would probably like to omit having to write the entire name of the module and then one of its methods. That's why we import it using the `as np` suffix in the import call. This way, instead of having to write `numpy.arange(10)` we can write `np.arange(10)`. This can prove to be quite effective when writing long scripts.

Importing a script into another script works exactly the same way. Suppose we have a script called "calculations.py", and in there we have defined a number of functions for various calculations, like additions, subtractions, multiplications, and divisions. Such a script could look like Script 2.12. Defining one-liners for these simple calculations might not seem very practical. Notice though the last function where we first make sure the denominator is not 0, as this would raise a `ZeroDivisionError`. If the denominator is zero, we return nothing, which will result in the variable that will store the returned value to be of `NoneType`. Also note that when we call `return` in any place of a function, the function exits, so whatever is written below it, will be omitted. This way we can use this keyword in various places inside a function, depending on what we want to return in each case.

Script 2.12 Arithmetic functions to be imported in other scripts.

```
1 def add(a, b):
2     return a + b
3
4 def subtract(a, b):
```

```
 5      return a - b
 6
 7 def multiply(a, b)
 8      return a * b
 9
10 def divide(a, b):
11      if b == 0:
12          return
13          return a/b
```

Suppose that we use this script so often that we want to import this script to another script, whenever we need any of these methods. This can be done just like importing any Python module, using `import`. Script 2.13 is an example.

Script 2.13 Importing our calculations.py script.
```
1 import calculations as cl
2 a = cl.divide(1, 0)
3
4 if a is None:
5     print("a is None, probably due to division by 0")
6 else:
7     print(f"a is {a}")
```

We imported our script giving it a short name, and then we called the division method, providing a 0 denominator. This way, Python will not crash, since we test against the denominator and return nothing in case of being 0. In this case, the `divide()` function will return a `NoneType` which we can test against and print some information as to what actually happened.

2.9.1 *Importing All Methods and Classes of a Module*

If we want to import all methods and classes of a module, but we don't want to prepend the name of the module every time we use one of its methods or classes, we can import a module or a script in another way.

Ingredients:

- Python3
- A text editor (preferably an IDE with Python support)
- A terminal window in case your editor does not launch Python scripts

Process:

Similarly to the way we imported the sleep method from the time module, we can import all the methods or classes of a module by using the asterisk instead of the name of a method. To import all the classes of the Pyo module, we must type the line in Shell 2.35.

Shell 2.35 Importing everything from the Pyo module.
```
>>> from pyo import *
```

The asterisking means "everything", and this way we import everything that is included in the Pyo module. Once we import a module like this, when we want to use one of its methods or classes, we write the name of the method or class directly, without prepending the name of the module. So, to use the `Sine()` class of the Pyo module, instead of writing the code in Shell 2.36, we can simply write the code in Shell 2.37.

Shell 2.36 Creating a sine wave oscillator by prepending the module name.
```
>>> a = pyo.Sine()
```

Shell 2.37 Creating a sine wave oscillator without prepending the module name.
```
>>> a = Sine()
```

Importing modules this way is considered to be bad practice because it is possible to get a conflict between names of methods or classes of two or more different modules, plus, we cannot know whether a class is native to Python or whether it comes from an external module. In computer programming jargon, this is called "polluting your namespace". If we import Pyo and one more module this way, and both modules have a class called `Sine()`, then it will not be clear which of the two modules is used. Pyo though, is focused on audio processing, and there are very few modules in Python that process audio, none of which is native to Python. Also, for the greatest part of this book, we will not use any module that has conflicting names with Pyo, so we will be mainly importing Pyo this way.

2.9.2   if __name__ == "__main__"

Before we move on to the next section, we will see what the if __name__ == "__main__" statement means.

Ingredients:

- Python3
- A text editor (preferably an IDE with Python support)
- A terminal window in case your editor does not launch Python scripts

Process:

Suppose that we have written a script that we sometimes want to use it as our main script, but other times we want to import it to another script. In this case, this statement, that is standard in Python, comes in handy. As an example, we will use our "calculations.py" script combined with the division by zero example. A modified version is shown in Script 2.14.

Script 2.14 Adding if __name__ == "__main__" to our calculations.py script.
```
1 def add(a, b):
2     return a + b
3
4 def subtract(a, b):
5     return a - b
6
7 def multiply(a, b)
```

```
 8        return a * b
 9
10 def divide(a, b):
11     if b == 0:
12          return
13     return a/b
14
15 def main():
16     a = divide(1, 0)
17
18     if a is None:
19         print("a is None, probably due to division by 0")
20     else:
21         print(f"a is {a}")
22
23 if __name__ == "__main__":
24     main()
```

Python includes a special variable for every script, called __name__ (two underscores before and after the word name). The string "__main__" is assigned to this variable only when a script is run as the main script. When a script is imported inside another script, then the actual name of the imported script is assigned to its __name__ variable. If we import the "calculations.py" script inside another script, then the string "calculations" will be assigned to its __name__ variable instead. This way we can test against the name of the __name__ variable – note the double equals signs in line 23 – and make sure that the main() function will not be called when we import "calculations.py" inside another script, but it will be called when we run it as the main script.

This feature is helpful for testing our script while developing it, before we load it in another script. In Script 2.14 we define a function to test if the 0 denominator test works as expected. When we run this script, we can determine whether our script runs properly, and if it doesn't, we can further develop it, until it works as expected. Then we can import this script in other scripts, and its main() function will not be called.

The word "main" was used in many places in the discussion of this example, and it might create some confusion. Note that the name of the function to be called if "calculations.py" is run as the main script does not have to be called main(). The naming of the function could be anything else, like func(), in which case, the last two lines of Script 2.14 should be changed to the two lines in Script 2.15.

Script 2.15 Changing the name of the main function.
```
23 if __name__ == "__main__":
24     func()
```

Naming the main function of a program main() though is a convention in Python. To better understand how the __name__ variable is used, write the code in Script 2.16 and save it with the name "my_script.py":

Script 2.16 Contents of my_script.py.
```
1 def print_name():
2     print(__name__)
```

```
3
4 if __name__ == "__main__":
5     print_name()
```

Now open a terminal window from the directory where this script is located and run it. You should see a __main__ printed on the console. Now type `python` to open the Python interpreter, and write the first two lines of Shell 2.38.

Shell 2.38 Importing my_script.py and printing its __name__ variable.

```
>>> import my_script as ms
>>> ms.print_name()
my_script
```

The third line will be printed by Python and you should not type it (note that there is no prompt at the beginning of this line). This is the string assigned to the __name__ variable of the "my_script.py" script, since it was not run directly, but it was imported.

2.10   Hello Sound!

After having built a foundation for programming in Python, we can finally produce some sound!

Ingredients:

- Python3
- A text editor (preferably an IDE with Python support)
- The Pyo module
- A terminal window in case your editor does not launch Python scripts

Process:

We have already seen how we can import modules in a Python script. Now we will import the Pyo module and use it. The code is shown in Script 2.17. We import all classes of the Pyo module using the asterisk, and the `sleep()` function from the time module. Python scripts, like programs written in other languages, run from top to bottom and then exit. If we do not include some way to stall our program for some time, we will not hear any sound at all, since our program will exit almost immediately. For this reason, we stall our program for five seconds using `sleep(5)`. Note that Pyo objects run on a different thread than the main Python thread, so stalling our program with `sleep(5)` has no effect on the sound.

When using Pyo, we must first create, boot, and start an audio server. This is done in lines 4 and 5. Line 4 creates and boots the server, and line 5 starts it. We create an object of the class `Server()` and call its `boot()` method on initialisation. This is a necessary step, before we create any other Pyo object. Then we call the `start()` method, so we can start the server and be able to get sound from Pyo objects. A method is a function that is inherent in a class. We create an object of the class `Sine()`, to get a sine wave tone at 440 Hertz (Hz).

Script 2.17 Outputting a 440Hz sine wave for five seconds.

```
1 from pyo import *
2 from time import sleep
```

```
 3
 4 s = Server().boot()
 5 s.start()
 6
 7 a = Sine(freq=440, mul=0.2).out()
 8
 9 def main():
10     sleep(5)
11
12 if __name__ == "__main__":
13     main()
```

Note that the `Sine()` class takes only kwargs. In this case we pass the `freq` and `mul` arguments. `freq` stands for "frequency", and `mul` for multiply. Apart from arguments to the class, `freq` and `mul` are also attributes of the class, as they each hold a value for a member of a class. We will learn more about classes in Chapter 8, but for now we should become accustomed to the terminology. The frequency attribute will set the frequency of the sine wave – its pitch – and the multiply attribute will set its amplitude – its loudness.

The multiply value is actually a coefficient by which the output of the sine wave oscillator is multiplied. If we multiply by 1, then we will get the full amplitude of the sine wave. If we multiply by a value greater than 1, we will amplify it, but in digital audio, when we amplify a signal beyond its normal amplitude, we actually get clipping (more on this in the next chapter). Here we multiply by 0.2, so we can hear a soft sine wave. Finally, we call the `out()` method of the `Sine()` class, to output the audio to the sound card of our system. `out()` can take one positional argument, which is the channel index, starting from 0 with the first speaker of our system, the left speaker in a stereo setup. If we do not supply an argument, it is equivalent to giving a 0 argument.

2.10.1 *Choosing an Audio Backend*

Pyo's server uses the PortAudio backend by default, but that might not be the backend used by your system. For example, in Linux, ALSA or Jack are the audio backends used by most users.

Ingredients:

- Python3
- A text editor (preferably an IDE with Python support)
- The Pyo module
- A terminal window in case your editor does not launch Python scripts

Process:

It is possible to set which audio backend Pyo will use, by passing the `audio` kwarg to the `Server()` class. Pyo can use PortAudio, Jack, and CoreAudio. Line 4 of Script 2.17 can change to the line in Script 2.18 so we can use Jack as the audio backend.

Script 2.18 Choosing the Jack audio server.
```
4 s = Server(audio="jack").boot()
```

Even if you haven't started Jack, Pyo will ask Jack to start in the background, but it will not ask it to stop, so you will have to stop Jack yourself. The rest of the previous script can be used intact.

2.11 Hello GUI!

If you have installed wxPython on your system, then you can utilise the GUI provided by Pyo. This is a set of widgets embedded in most of its classes that allow for manual control over their parameters.

Ingredients:

- Python3
- A text editor (preferably an IDE with Python support)
- The Pyo module
- The wxPython module
- A terminal window in case your editor does not launch Python scripts

Process:

In this example we will see how flexible Python can be, and how Pyo takes advantage of Python's flexibility, to create complex structures with a handful of lines of code. Script 2.19, which is a modified version from the examples provided by the E-Pyo editor, creates a list of eight sine wave oscillators and binds them together in a widget that enables us to control the frequency of each oscillator separately.

Script 2.19 Using Pyo's GUI.
```
 1 from pyo import *
 2
 3 s = Server().boot()
 4 s.amp = 0.1
 5
 6 oscs = Sine([100, 200, 300, 400, 500, 600, 700, 800],
 7               mul=0.1).out()
 8 oscs.ctrl(title="Simple additive synthesis")
 9
10 s.gui(locals())
```

If you run this code, two windows will pop up, shown in Figures 2.1 and 2.2. Figure 2.1 shows the Pyo Server window where we can start and stop the audio processing, and Figure 2.2 shows the window titled "Simple additive synthesis" (the string argument passed to the `oscs.ctrl()` method), which enables us to control the frequency of each oscillator separately.

Note that the Pyo Server window includes a "Rec Start" button. If you click on this while the audio is running, a file named "pyo_rec.wav" will be saved in your home directory. This will be a recording what whatever Pyo is playing while this button is pressed. It is a convenient way to store your sounds. There is also a slider for the overall amplitude in Decibels (labelled "dB"),

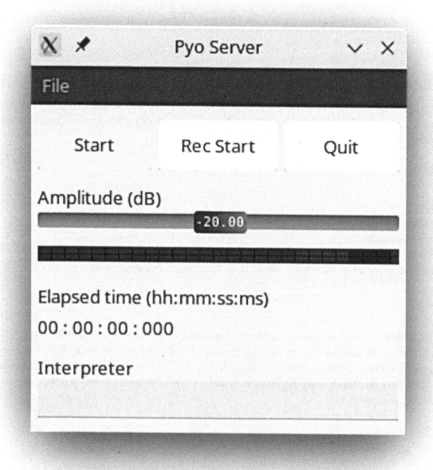

Figure 2.1 The Pyo Server window.

and a VU meter of the audio output. At the bottom there is a text field labelled "Interpreter". We will talk about this in the next section.

In the "Simple additive synthesis" window you can drag each slider from the multi-slider in the "freq" window and you will hear the frequency of the corresponding oscillator change. On the bottom part there are two more sliders, one for the phase and one for the amplitude, labelled "mul". The Sine() class has these three kwargs, freq, phase, and mul, so these sliders appear when you call its ctrl() method (there is one more argument, add, but no slider appears for that).

There are a couple of things to note in the code of this example. First of all, we set a value to the amp attribute of the Server() class. This will set the overall amplitude of the program, in contrast to the mul attribute of the Sine() class, which sets the amplitude of the sine wave oscillator only. The second thing is the ctrl() method of the Sine() class, which displays the "Simple additive synthesis" widget, and the last line of our code, which calls the gui() method of the Server() class. This method launches Pyo's GUI and displays the Server window. If we do not include this method call in our code, no GUI will be displayed, and our program will

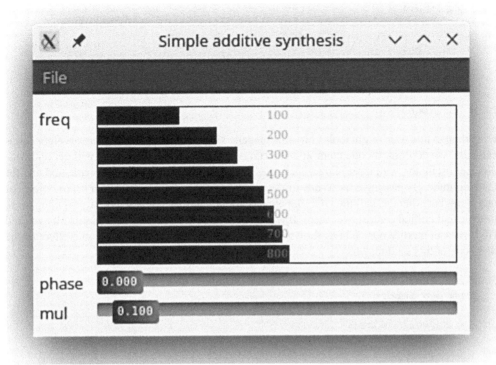

Figure 2.2 The widget that controls the frequencies of the eight oscillators.

exit immediately, since there is no mechanism to stall our program, as there was in the example of the previous section, where we called `sleep(5)`.

The `gui()` method of the `Server()` class, apart from enabling the GUI, also runs a loop that keeps the Python program alive, without us having to use `sleep()`, or any other mechanism. The `locals()` function call passed as an argument to `gui()` returns a dictionary of the current local symbol table. You can add a `print(locals())` line in your script, so you can see what this table is – it will print a long dictionary with names of variables and classes as keys, and their values or types as values.

Note that we could also write this code using list comprehension, the way it is done in Script 2.20. Line 6 of this script creates eight objects of the `Sine()` class with frequencies 100Hz apart, with the first one at 100Hz, the second at 200Hz, etc., exactly as they were hard-coded in the previous example. Notice the argument passed to the `out()` method, `i%2`. The percentage symbol in Python and other languages is called the modulo operator. This operator yields the remainder of the division between the two values on either side of it. In this case it is used to yield an alternating 0 and 1, as the variable `i` will increment from 0 to 8, excluding 8. This way we are sending the oscillator objects with even indexes to the left channel and with the odd indexes to the right.

Script 2.20 Creating eight oscillators with list comprehension.

```
1 from pyo import *
2
3 s = Server().boot()
```

```
 4 s.amp = 0.1
 5
 6 oscs = [Sine(freq=(i*100+100),mul=0.1).out(i%2)
 7          for i in range(8)]
 8 for i in range(8):
 9     oscs[i].ctrl(title=f"Simple additive synthesis {i}")
10
11 s.gui(locals())
```

Even though this code might look a bit more elegant, since we define the frequencies algorithmically and we don't hard-code them in our script, when the GUI launches, we will get one window per oscillator, with three sliders each, one for the frequency, one for the phase, and one for the amplitude. By passing a list as any of the kwargs of a Pyo class, we create a list of objects of this class, but when we call the GUI, the objects are bound to a single window, providing more flexible and unified control over them. If we want to combine list comprehension with a unified GUI, we can modify our code as shown in Script 2.21 so it has exactly the same effect as the first code of this section.

Script 2.21 Alternative use of list comprehension that unifies the GUI.

```
 1 from pyo import *
 2
 3 s = Server().boot()
 4 s.amp = 0.1
 5
 6 freqs = [(i*100+100) for i in range(8)]
 7 oscs = Sine(freq=freqs, mul=0.1).out()
 8 oscs.ctrl(title="Simple additive synthesis")
 9
10 s.gui(locals())
```

2.12   Combining Interpreter Input with Predefined Scripts

In this section we will see how we can combine a predefined script with input from the interpreter. This is possible in a few different ways, but we will look at how we can achieve this with Pyo's GUI.

Ingredients:

- Python3
- A text editor (preferably an IDE with Python support)
- The Pyo module
- The wxPython module
- A terminal window in case your editor does not launch Python scripts

Process:

Pyo's GUI also supports giving interpreter input to a running script by providing a text entry in the Server window. Script 2.22 creates a Frequency Modulation (FM) effect. We will see FM in detail in the next chapter, so for now we will only focus on the combination of scripts with live input. All we need to know is that the sine wave oscillator named `car` is the carrier oscillator, and the one named `mod` is the modulator.

Script 2.22 A script to be run interactively with Pyo's GUI.
```
1 from pyo import *
2 from time import sleep
3
4 s = Server().boot()
5
6 mod = Sine(freq=205,mul=500,add=200)
7 car = Sine(freq=mod,mul=.2).out()
8
9 s.gui(locals())
```

Figure 2.3 shows the server window with input to change the frequency of mod, at the bottom part of the window. This entry is labelled "Interpreter" and it is exactly that, the Python interpreter where you can write Python one-liners. Experiment with input for other attributes, like `mul` and `add` for both oscillators.

Figure 2.3 The Pyo Server window with interpreter input.

2.13   Conclusion

We have covered a lot of ground in this chapter and now we are able to move on and learn the specifics of digital audio and start making music applications. We have developed a foundation for programming with Python. We learnt about the different data types, conditional tests, functions, loops, the scope of variables, how to import modules, how to create sound, and more Pyo specifics, like Pyo's GUI and how to combine scripts with live input. The only thing that this chapter did not cover is how to write our own classes in Python. This will be covered in Chapter 8, where we will create classes that will include features that are not present in the native classes of Pyo.

Even though this chapter provided all the necessary information to start programming in Python on your own, keep in mind that computer programming needs practice, and one becomes better at it only through systematic and focused practice. The next section includes some exercises based on what we have learnt so far. You are strongly encouraged to cover this section and pursuit your own way into programming. There are many online resources with tutorials or forums where one can find answers. Looking up those resources before asking a question is considered to be good practice and a standard in netiquette. After all, forums and mailing lists exist so that questions and answers are documented, and people can look those up in the future, without needing to ask the same questions again.

2.14   Exercises

2.14.1   Exercise 1

Write a function that takes one argument and prints its type on the console.

2.14.2   Exercise 2

Write a loop using both `for` and `while` that iterates over a list of even numbers from 0 to 100 and skips iterations at the beginning of every ten, e.g. 0, 10, 20, etc.

Tip: when using `range()`, provide three arguments, a start value, a stop value, and a step. To get the number 0 out of any number in your list, use the modulo operator. Note that you cannot provide 0 as the right-hand operand to modulo because it will result in a `ZeroDivisionError`. To determine whether your loop works as expected, use `print()`.

2.14.3   Exercise 3

Take the first example from the "Hello GUI!" section and add a function that takes two arguments, an index and a value. This function should change the frequency of the oscillator at the given index to the given value.

Tip: in this code, `oscs` is a list, so index it as you would index a list. Every item in this list is an object of the `Sine()` class. This class has a method `setFreq(x)` for setting the frequency, where x is a float (or another Pyo object, but more on that in the next chapter). Use the Interpreter entry field in the Server window of Pyo's GUI to call your function. Note that once you change the frequency of an oscillator this way, it is not possible to control it from the GUI any more.

3 Diving into Audio Synthesis Techniques

In this chapter we will focus on audio synthesis techniques that we will realise with Pyo. We will cover basic techniques and concepts in digital audio, and we will discuss about certain intricacies of this domain. The approach taken in this chapter does not require a mathematical background. For the more maths-inclined readers, more details on certain techniques, like Ring Modulation, Frequency Modulation, Fast Fourier Transform, and others, can be found in the legendary *Computer Music Tutorial* by Curtis Roads. Even if you are advanced in digital audio, you are encouraged to read through this chapter, as we will be focusing on the way Pyo creates sound through its various classes and their methods.

What You Will Learn

After you have read this chapter you will:

- Know about oscillators and waveforms, the basic elements of electronic sound
- Be acquainted with standard digital audio synthesis techniques
- Be acquainted with various audio effects
- Know how to control timbral and temporal sound parameters
- Know about sampling rate, buffer size, and audio range and clipping
- Know about the basics of the Fast Fourier Transform

3.1 Oscillators and Waveforms

The most fundamental element of electronic sound is the oscillator. This is a sound generator that produces a tone with a certain timbre at a specified pitch. The timbre of an oscillator is determined by its waveform. Its pitch is determined by the number of repetitions of this waveform, referred to as its frequency. Frequency is measured in Hertz (Hz), where this unit represents the number of repetitions per second. The amount of time an oscillator takes for one such repetition is called a period (Farnell, 2010, p. 39), or a cycle.

The fundamental waveforms are four: the sine wave, the triangle, the sawtooth, and the square wave. These are shown in Figures 3.1 to 3.4. All waveforms take their names from their shape. Oscillators are used both for audio and to control other parameters of sound. In this case, their frequency is usually in an infra-audio range. For this reason, this type of oscillator is called a Low Frequency Oscillator, or an LFO. In Pyo (and digital audio processing in general), an oscillator can be either an audible oscillator or an LFO.

DOI: 10.4324/9781003386964-3

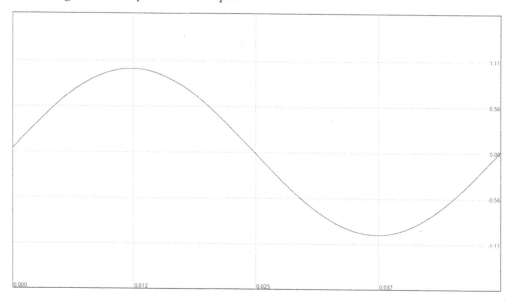

Figure 3.1 A sine wave.

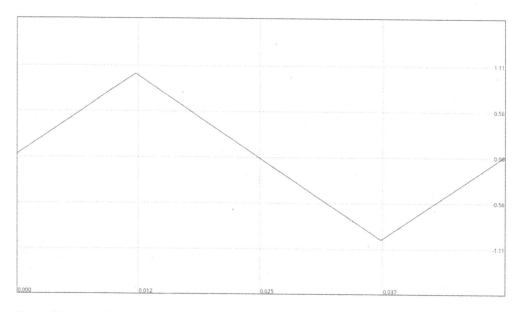

Figure 3.2 A triangle wave.

A sound is usually made out of a complex structure. Every periodic sound that we perceive it to have a definite pitch is produced by a series of pure tones – sine waves – where usually the fundamental – the tone with the pitch we perceive – has the greatest amplitude, and the rest have a higher pitch and lower amplitude. These tones are called harmonics or partials. All partials, except the fundamental, are called overtones. In harmonic sounds, the frequencies of the overtones

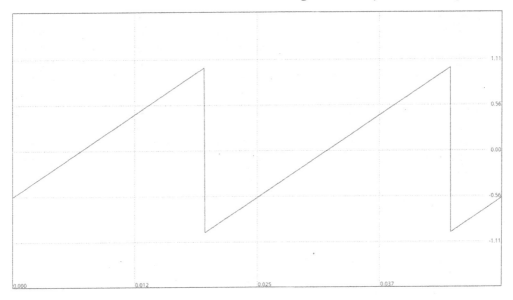

Figure 3.3 A sawtooth wave.

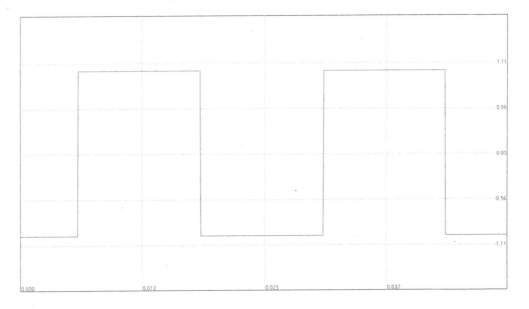

Figure 3.4 A square wave.

are integer multiples of the fundamental (Roads, 1996, p. 16). A complex harmonic tone at 400Hz has a fundamental at 400Hz, the first overtone at 800Hz, the second overtone at 1200Hz, and so on. Non-harmonic sounds – sounds that we perceive as noise – have non-discernible patterns in their overtone structure (Roads, 1996, p. 14).

The sine wave is the purest tone possible, as it includes a single harmonic – the fundamental – and no overtones. The triangle comes next with a few overtones. Their amplitude is defined as $1/N^2$, where N is an odd number and represents the number of the partial. Even partials have an amplitude of 0. The amplitude of every other odd partial has an inverted sign, which means that the amplitude of the first partial is positive, the amplitude of the third partial is negative, of the fifth is positive, and so on. The sawtooth waveform includes both odd and even partials, and their amplitude is defined as $1/N$, where N is the number of the partial. The square wave is a combination of the triangle and the sawtooth, as the amplitude of its partials is defined as $1/N$, where N is the partial number, but again it includes only odd partials, all with positive amplitudes, with even partials having an amplitude of 0. Spectral analyses of all four waveforms are shown in Figures 3.5 to 3.8.

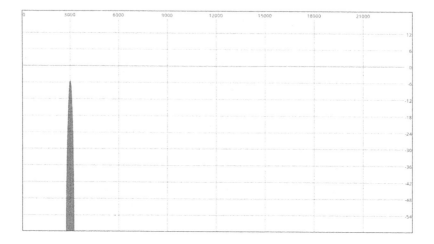

Figure 3.5 Spectrum of a sine wave at 3,000Hz.

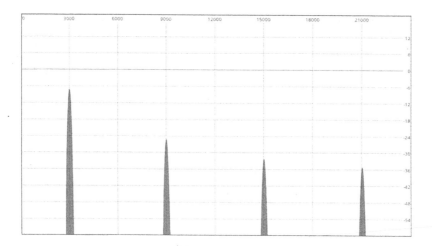

Figure 3.6 Spectrum of a triangle wave at 3,000Hz. Only odd partials are present.

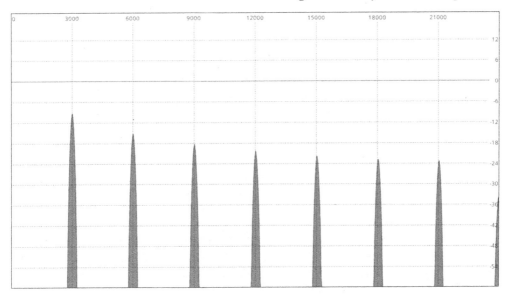

Figure 3.7 Spectrum of a sawtooth wave at 3,000Hz. Both odd and even partials are present.

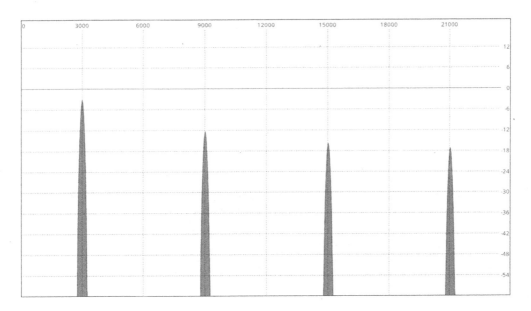

Figure 3.8 Spectrum of a square wave at 3,000Hz. Only odd partials are present.

Ingredients:

- Python3
- A text editor (preferably an IDE with Python support)
- The Pyo module

- the wxPython module, if you want to display waveforms
- A terminal window in case your editor does not launch Python scripts

Process:

Pyo includes classes for many waveform, including the four standard ones. As we have already seen, the sine wave is produced with the Sine() class. The triangle is produced by two classes, RCOsc() and LFO(). The sawtooth is produced by Phasor(), although Phasor() is actually a ramp from 0 to 1, whereas the full amplitude of a digital audio signal has a range between -1 and 1 – more on audio range at the end of this chapter. The square wave can be produced with Phasor() and a few other classes. LFO() and RCOsc() can produce a quasi-square oscillator.

We have already seen the Sine() class in Chapter 2, so we will skip this oscillator. Script 3.1 produces a triangle and a sawtooth oscillator, and we can choose which one to display in the oscilloscope with a Selector() object. To get a triangle we must set the type attribute of LFO() to 3. To map the output of Phasor() to the full range we multiply it by 2 and subtract 1. To change the oscillator displayed in the oscilloscope, in the interpreter entry, type sel.setVoice(1) for the sawtooth, and sel.setVoice(0) for the triangle.

Script 3.1 A triangle and a sawtooth oscillator.
```
 1 from pyo import *
 2
 3 s = Server().boot()
 4
 5 tri = LFO(freq=440, sharp=1, type=3)
 6 saw = Phasor(freq=440, mul=2, add=-1)
 7 sel = Selector(inputs=[tri, saw], voice=0.0)
 8
 9 sc = Scope(sel)
10
11 s.gui(locals())
```

Even though LFO() and RCOsc() produce a quasi-square wave oscillator, to create a square wave like the one in Figure 3.4, we must combine a few Pyo classes. The code in Script 3.2 creates a square wave oscillator and displays it through Pyo's oscilloscope.

Script 3.2 A square wave oscillator.
```
1 from pyo import *
2
3 s = Server().boot()
4
5 a = Phasor(freq=440)
6 b = Round(a, mul=2, add=-1)
7 sc = Scope(b, gain=1)
8
9 s.gui(locals())
```

Round() takes an input and rounds it to the closest integer. In the code above the Phasor() goes from 0 to 1 at a frequency of 440Hz, so during the first half of its period, Round() will round its output to 0, and during the second half it will round it to 1. If we multiply this output by 2 and subtract 1, through the `mul` and `add` kwargs, we bring the output to the full audio range, between -1 and 1.

If you want to output these waveforms to your speakers, make sure you turn the volume down, otherwise you will probably get a loud sound. Sending the signal to your computer speakers is done the same way we created sound in the "Hello Sound!" section in the previous chapter.

There is one more thing to point out, before we move on to the next section. In all the examples above we called `s.gui(locals())` in the last line of our code. To be able to use Pyo's GUI you will need to have wxPython installed in your system. Chapter 1 covers the installation process of this Python module. If you don't have wxPython, instead of this line you should write `s.start()`. You should also include a loop to keep your script alive, the same way we did in the "Hello Sound!" section in the previous chapter, unless you run your code straight in a Python interpreter, either in IDLE, a Jupyter notebook, or in a terminal window. Note though that without wxPython, you cannot display the waveforms, so the lines that create a `Scope()` object will not work, and you will probably get an error message. In this case, remove every line that creates a `Scope()` object. Throughout this book it will be assumed that you have wxPython, so the code will be written with this module in mind. You will still be able to use this code without this module, but you will have to apply these small changes.

Documentation Pages:

`Server()`: https://belangeo.github.io/pyo/api/classes/server.html#server
`LFO()`: https://belangeo.github.io/pyo/api/classes/generators.html#lfo
`Phasor()`: https://belangeo.github.io/pyo/api/classes/generators.html#phasor
`Selector()`: https://belangeo.github.io/pyo/api/classes/pan.html#selector
`Scope()`: https://belangeo.github.io/pyo/api/classes/analysis.html#scope
`Round()`: https://belangeo.github.io/pyo/api/classes/arithmetic.html#round

3.2   Audio Synthesis

Now that we have covered the fundamental elements of electronic sound, we can move on to more musical processes. In this section we will cover some of the most standard synthesis techniques in electronic music, including Ring Modulation, Amplitude Modulation, Frequency Modulation, Additive Synthesis, and Granular Synthesis.

3.2.1   *Ring Modulation*

The first synthesis technique we will look at is the Ring Modulation, abbreviated RM. This technique takes its name from its analog implementation, where four diodes are arranged in a "ring" configuration (Roads, 1996, p. 220).

Ingredients:

- Python3
- A text editor (preferably an IDE with Python support)
- The Pyo module
- the wxPython module, if you want to display waveforms
- A terminal window in case your editor does not launch Python scripts

Process:

Digital RM is a simple multiplication of the output of two oscillators, and there are two ways we can achieve this in Pyo. The first one is shown in Script 3.3.

Script 3.3 A Ring Modulator.
```
 1 from pyo import *
 2
 3 s = Server().boot()
 4
 5 mod = Sine(freq=400)
 6 car = Sine(freq=1000)
 7 rm = Sig(car*mod, mul=.2).out()
 8
 9 sp = Spectrum(rm)
10 sc = Scope(rm)
11
12 s.gui(locals())
```

In the code of Script 3.3 we have created two sine wave oscillators and we multiply them in the input entry of the `Sig()` object. The `Sig()` class creates a signal at audio rate from arbitrary values and it is very convenient when we want to output the result of such calculations to our speakers. The `mul` kwarg is set to 0.2 (note that in Python we can omit the 0 and write .2, which is the same as 0.2) so we don't get a loud sound when we run our code. We have also created a `Spectrum()` and a `Scope()` object so we can inspect the waveform and the spectrum of the result.

The names given to the objects in lines 5, 6, and 7 are taken from the terminology used in audio synthesis techniques. The first sine wave is called mod because it is the modulator. The second is called `car` because it is the carrier. The `Sig()` object is called `rm` because it is the RM (Ring Modulation) output. In the case of RM, exchanging the modulator with the carrier has no effect, as the order of the operands in multiplication is insignificant. It is a good idea though to get used to them as their role is more effective in the other synthesis techniques we will see further on.

The result of RM is two tones with distinct pitches, where one tone will have the pitch of the sum of the frequencies of the carrier and the modulator, and the other tone will have the pitch of their difference. These are called sidebands. In the example above, one sideband will have a frequency of 1400Hz and the other will have a frequency of 600Hz.

In this example, the resulting sound is coming out of one speaker only, the left one in a stereo set up. To output the sound in stereo, change line 7 of the code of Script 3.3 to the line of Script 3.4 This will create a `Mix()` object that will take the multiplication of the carrier and modulator oscillators as its input signal, and by setting the `voices` attribute to 2, it will output the signal in stereo.

Script 3.4 Stereo output of the Ring Modulator.
```
7 rm = Mix(car*mod, voices=2, mul=.2).out()
```

Note that a sine wave is not the only waveform we can use. Instead of a sine wave we can use any waveform we want. We can even use a microphone input for the carrier and modulate that with a sine wave. To do that, replace the `Sine()` class of the carrier with the `Input()` class with no arguments, or set a value in its `mul` attribute to a value lower than 1 to avoid feedback, in case the microphone is close to the speakers.

Another way to achieve RM in Pyo is by setting the modulator to be the argument passed to the `mul` attribute of the carrier oscillator. The code in Script 3.5, which is very similar to the one of Script 3.3, shows this.

Script 3.5 Another version of the Ring Modulator.

```
 1 from pyo import *
 2
 3 s = Server().boot()
 4
 5 mod = Sine(freq=400)
 6 car = Sine(freq=1000, mul=mod)
 7 rm = Sig(car, mul=.2).mix(2).out()
 8
 9 sp = Spectrum(rm)
10 sc = Scope(rm)
11
12 s.gui(locals())
```

The only difference between the two versions of RM is in lines 6 and 7. Instead of multiplying the two oscillators in the input entry of the `Sig()` object, we pass the modulator oscillator as an argument to the `mul` attribute of the carrier, in line 6. In line 7 we pass only the carrier in the input entry of the `Sig()` object. In this case, we can send the carrier oscillator straight to the output, but we will not have any control over its amplitude, as the `mul` attribute of the carrier is controlled by the modulator. For this reason, we pass our ring-modulated signal to a `Sig()` object where we can control the amplitude of the RM output. In line 7, we call the `mix()` attribute, to turn the signal to stereo. This implicitly switches the `rm` object from an instance of the `Sig()` class to one of `Mix()`.

Figure 3.9 illustrates the process of this last version of RM. This illustration depicts two oscillators with two inputs each, one for the frequency and one for the amplitude. The modulator receives arbitrary inputs, while the carrier can have any frequency, but its amplitude is controlled by the output of the modulator. Note that both oscillators have the full audio range. This is the only difference between Ring Modulation and Amplitude Modulation, which we will see next.

Documentation Pages:

`Sine()`: https://belangeo.github.io/pyo/api/classes/generators.html#sine
`Spectrum()`: https://belangeo.github.io/pyo/api/classes/analysis.html#spectrum
`Mix()`: https://belangeo.github.io/pyo/api/classes/internals.html#mix

3.2.2 Amplitude Modulation

The next synthesis technique we will look at is Amplitude Modulation, abbreviated AM. This is a technique where the amplitude of one oscillator, called the carrier, is modulated by another oscillator, called the modulator.

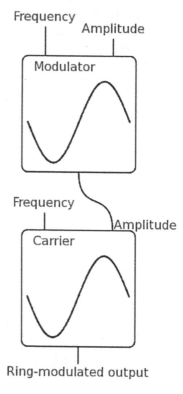

Figure 3.9 Ring Modulation process. The output of the modulator controls the amplitude of the carrier.

Ingredients:

- Python3
- A text editor (preferably an IDE with Python support)
- The Pyo module
- the wxPython module, if you want to display waveforms
- A terminal window in case your editor does not launch Python scripts

Process:

AM is very similar to RM, only in AM, the modulator is unipolar, meaning that its output is in the range between 0 and 1, instead of being between -1 and 1. AM is used to create effects like tremolo, where the amplitude of the carrier oscillates in a fast rate, yet in the infra-audio range. The code in Script 3.6 is similar to the second version of RM, only the modulator is now unipolar:

Script 3.6 An Amplitude Modulator.
```
1 from pyo import *
2
3 s = Server().boot()
4
```

```
 5 mod = Sine(freq=15, mul=0.5, add=0.5)
 6 car = Sine(freq=400, mul=mod)
 7 am = Mix(car, voices=2, mul=.2).out()
 8
 9 sp = Spectrum(am)
10 sc = Scope(am)
11
12 s.gui(locals())
```

To make the modulator unipolar we need to multiply its output by 0.5 and add 0.5. The `mul` attribute scales the output, and the `add` attribute gives an offset to it. In this case, by multiplying by 0.5, we scale the range of the oscillator from a range from -1 to 1, to a range from -0.5 to 0.5. By adding 0.5, we offset this range to now go from 0 to 1. Note that the frequency of the modulator is low. A 15Hz frequency is not audible by the human ear, so if you output this oscillator to your speakers (even if you have a sound system that can play such low frequencies), you will not hear anything. The effect of AM though, is perfectly audible if we set the frequency of the modulator to this infra-audio range.

If we raise the frequency of the modulator and bring it in the audio range, we will start hearing an effect similar to RM. The difference between RM and AM is that in AM, alongside the sum and difference of the carrier and modulator frequencies, we can also hear the carrier frequency. If we change the modulator frequency to 1000Hz, like we did with the RM example, we will hear three tones, one at 600Hz (the difference), one at 1000Hz (the carrier), and one at 1400Hz (the sum). Figure 3.9 depicts the process of both RM and AM, with the only difference being the range of the modulator, where in RM it is between -1 and 1, and in AM it is between 0 and 1.

In electronic music, AM is used very frequently, but most of the times not to create a tremolo effect. Modulating the amplitude of a sound source is used to control its amplitude over time. This is usually done with envelopes, which we will see later on in this chapter.

3.2.3 *Frequency Modulation*

One of the most celebrated techniques in electronic music is the Frequency Modulation, abbreviated FM. It is somewhat similar to AM, as in FM, the modulator modulates the frequency of the carrier, instead of its amplitude, but it is a bit more involved than RM and AM, as it includes more elements.

Ingredients:

- Python3
- A text editor (preferably an IDE with Python support)
- The Pyo module
- The wxPython module, if you want to display waveforms
- A terminal window in case your editor does not launch Python scripts

Process:

Although FM has had many non-musical applications dating back to the nineteenth century (Roads, 1996, p. 225), John Chowning was the first to explore its musical potential (Chowning, 1973; Roads, 1996, p. 225). In FM, the output of the modulator is multiplied by a value called the "index", and added to the frequency of the carrier. Figure 3.10 illustrates this process.

One application of FM is the simulation of vibrato. If the frequency of the modulator and the value of the index are both low, then a slowly fluctuating value will be added to the frequency of the carrier. Note that the output of the modulator is bipolar. If the frequency of the modulator is 1Hz, the index is 2, and the frequency of the carrier is 440Hz, then we will hear the waveform of the carrier sliding up and down between 338Hz and 442Hz once per second. Since the output of the modulator is bipolar going from -1 to 1, and it is multiplied by 2, after this multiplication, the modulator signal goes from -2 to 2. When we add this signal to the 440Hz value of the carrier frequency, we get this small and slow fluctuation around 440Hz.

FM becomes more interesting though when we set high values for both the modulator frequency and the index. As the frequency of the modulator enters the audio range, what we perceive instead of a vibrato is a sound rich in harmonics, again called sidebands. The structure of these sidebands depends on the ratio of the modulator and carrier frequencies, and their amount depends on the value of the index. As this value raises, more sidebands start to appear on both sides of the carrier frequency. When the ratio of the carrier and modulator frequencies is an integer, then the resulting sound is harmonic. When this ratio is not an integer, the sound becomes inharmonic. The code in Script 3.7 implements FM, and Figure 3.11 illustrates the spectrum of it.

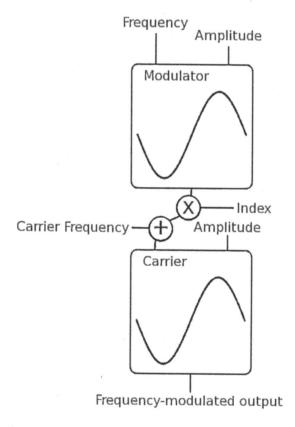

Figure 3.10 Frequency Modulation process. The output of the modulator is multiplied by the index and added to the frequency of the carrier.

Script 3.7 A Frequency Modulator.

```
1 from pyo import *
2
3 s = Server().boot()
4
5 ndx = Sig(150)
6 mod = Sine(freq=200, mul=ndx)
7 car = Sine(800+mod, mul=.2).mix(2).out()
8
9 sp = Spectrum(car, size=4096)
10
11 s.gui(locals())
```

In Script 3.7 we create a `Sig()` object that holds the value of the index, and we pass this to the `mul` attribute of the modulator oscillator. We then add a constant value of 800 – which is the frequency of the carrier – to the modulator in the frequency entry of our carrier oscillator. We finally set the `mul` attribute to a low value to get a soft sound, and we mix the output to both channels of a stereo setup. In line 9 we create a `Spectrum()` object and we pass the value 4096 to its `size` attribute.[1] This is the size of the window of the Fast Fourier Transform analysis. We will see more about Fast Fourier Transform later on in this chapter.

In the spectral analysis shown in Figure 3.11, we can see that the most prominent frequency is at 800Hz, which is the frequency of the carrier, and then we have decaying frequencies at 1000Hz, and 1200Hz – which are the sum of the carrier and integer multiples of the modulator, 800+200 and 800+400 – and at 600Hz and 400Hz – which are the differences of the carrier and integer multiples of the modulator, 800–200 and 800–400.

Since FM is so popular in electronic music, Pyo includes a class especially for this. By using this class we can compress lines 5 to 7 of our code to one line. This is shown in Script 3.8.

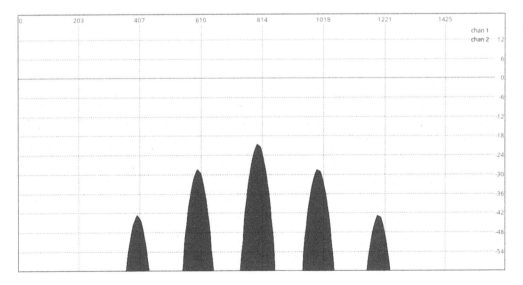

Figure 3.11 Spectrum of FM with a carrier frequency of 800Hz, modulator frequency of 200Hz (4:1 ratio), and index of 150.

Script 3.8 Frequency Modulation with the FM() class.

```
1 from pyo import *
2
3 s = Server().boot()
4
5 fm = FM(carrier=800,ratio=0.25,index=0.75,mul=.2).mix(2).out()
6
7 sp = Spectrum(fm, size=4096)
8
9 s.gui(locals())
```

Now line 5 contains an object of the FM() class. The result of this code is identical to the result of the previous one. The kwargs of the FM() class are the carrier frequency, the ratio of the carrier and the modulator, the index which is multiplied by the modulator frequency and gives the amplitude of the modulator, and the mul and add attributes which are common between most of the Pyo classes. Since we provide a ratio of 0.25, we get a frequency of 200Hz for the modulator. Multiplying this by 0.75 yields 150, which is the index we used in the previous example.

We can also animate our sound by modulating the index. To do this, we can set the index to be an oscillator, and set its range through the mul and add attributes. We can change our code to the one in Script 3.9.

Script 3.9 Frequency Modulation with an animated index.

```
1 from pyo import *
2
3 s = Server().boot()
4
5 ndx = Sine(freq=0.5, mul=2, add=3)
6 fm = FM(carrier=800,ratio=0.25,index=ndx,mul=.2).mix(2).out()
7
8 sp = Spectrum(fm, size=4096)
9 s.gui(locals())
```

Line 5 creates a sine wave oscillator with a frequency of 0.5Hz that goes from 1 to 5 – since the Sine() class outputs a bipolar signal from -1 to 1, we multiply by 2 to scale it from -2 to 2, and add 3 to offset it to go from 1 to 5. We now pass this sine wave oscillator object to the index attribute of the FM() class and we hear a sweeping spectrum that goes from bright to damper and back, in a sine wave fashion.

Last thing to note is that the FM() class uses sine waves. If you want to do FM with other waveforms, you will have to use the first example, where we modulated the frequency of the carrier explicitly by using separate oscillators.

Documentation Page:

FM() : https://belangeo.github.io/pyo/api/classes/generators.html#fm

3.2.3.1   Phase Modulation

We can get the same results we get with FM if we modulate the phase of an oscillator instead of its frequency. This technique is called Phase Modulation, abbreviated PM. The example of Script 3.7 can be replaced with the code in Script 3.10.

Script 3.10 A Phase Modulator.

```
1 from pyo import *
2
3 s = Server().boot()
4
5 mod = Sine(freq=200, mul=.1, add=.4)
6 car = Sine(800, phase=mod, mul=.2).mix(2).out()
7
8 sp = Spectrum(car, size=4096)
9
10 s.gui(locals())
```

If you see the spectrum of the output sound you will see it is identical to the spectrum of the FM example. In this case, the output of the modulator is squeezed to a range between 0.3 and 0.4. In PM, the output of the modulator must be positive, so the index has been integrated in the `mul` and `add` attributes of the modulator. You should also make sure that the output of the modulator does not exceed 1, or the output of the carrier will reach very high values.

3.2.4 Additive Synthesis

Additive synthesis is a technique of adding together multiple sine waves to create a more complex waveform. The concept of additive synthesis is centuries old as it can be found in the register-stops of pipe organs (Roads, 1996, p. 134). These register-stops open up or close groups of pipes that let air through when open, thus forming the timbre of the organ sound. The function of additive synthesis can also be found in any natural sound, as sound is a set of complex oscillations produced simultaneously (Cipriani and Giri, 2009, p. 187). In this section we will see how we can create the standard waveforms by means of additive synthesis, as well as other, arbitrary waveforms.

Ingredients:

- Python3
- A text editor (preferably an IDE with Python support)
- The Pyo module
- the wxPython module, if you want to display waveforms
- A terminal window in case your editor does not launch Python scripts

Process:

The first waveform we will create by means of additive synthesis is the sawtooth, because it is the simplest to implement. The code in Script 3.11 creates what is called a band-limited sawtooth with 50 harmonics. Remember that a sawtooth is created by a set of partials whose frequencies are integer multiples of the fundamental, and their amplitude is defined as 1/N, where N is the number of the partial. In lines 7 and 8 we do just that. We create a list of sine waves using list comprehension. Line 5 defines a base frequency, and in the list comprehension we multiply this variable by the index `i`. Note the arguments passed to `range()` in the list comprehension. These are 1 and 51. This means that `range()` will return a list where the first value will be 1, the last will be 50 (51 is excluded), and since we don't provide a third argument, the values of the list will increment by 1. The `for` loop will iterate over this list and will assign each value to `i`. This way, the value passed to the freq attribute of the `Sine()` class will result in integer multiples of the base frequency, since `i` will be 1 in the first iteration, 2 in the second, 3 in the third, and so on. The value passed to the `mul` attribute is the inverse of the value of `i`, since we divide 1 by `i`, which is essentially the same as 1/N, as we have already defined as the formula for setting the amplitudes of the partials of the sawtooth.

Script 3.11 A band-limited sawtooth oscillator via Additive Synthesis.

```
1 from pyo import *
2
3 s = Server().boot()
4
5 base_freq = 100
6
7 sines = [Sine(freq=(i*base_freq), mul=(1/i))
8          for i in range(1,51)]
9 mixer = Mixer(outs=1, chnls=2)
10
11 for i in range(50):
12     mixer.addInput("additive"+str(i), sines[i])
13     mixer.setAmp("additive"+str(i), 0, .5)
14
15 sig = Sig(mixer[0], mul=.5).out()
16
17 sc = Scope(sig, gain=1)
18 sp = Spectrum(sig)
19
20 s.gui(locals())
```

Line 9 needs some explanation, as to what it does, and why we use it. Suppose we omit this line and we output the sine waves by calling their `out(i%2)` method, as we have done in Chapter 2. This will split the sine waves in half, and will send the odd to the left speaker, and the even to the right. This will also result in the left speaker outputting a square wave – since only odd harmonics will be sent there – and the right speaker outputting a sawtooth, with the right speaker having double the base frequency as a fundamental. If you want to test this, make sure you set the volume of your sound system low.

If we pass the sines list – which contains 50 objects of the `Sine()` class – as the value to a `Mix()` object, we will again get the same stereo output, with the sine waves split to two groups. To tackle this issue we use a `Mixer()` object where we mix all the sine waves to a single signal. We provide the following arguments to the `Mixer()`, one output through the `outs` kwarg and two channels per output through the `chnls` kwarg to get a stereo output. Since we want to mix the 50 sine waves down to one signal, we need our mixer to have only one output. The `chnls` attribute determines how many channels each of the outputs of the mixer will have. In this example we set this value to 2, since we want stereo.

In lines 11 to 13 we have a for loop where we add inputs to our mixer object and we set amplitude values. The loop runs 50 times, since we have 50 partials in our signal. Line 12 adds an input to the mixer. The first argument of the `addInput()` method is a dictionary key. To make these keys unique we concatenate the string "additive" with the value of `i` converted to a string using the built-in Python `str()` class. This will result in the strings "additive0", "additive1", "additive2" etc., until "additive49".

The second argument is the value of the key in the mixer's dictionary, in our case, each of the `Sine()` objects of the `sines` list. In line 13 we set an amplitude value for each value of the mixer's dictionary, and we set which output it will be routed to. The first argument is the dictionary key, the second is the output index, and the third is the amplitude value. By calling the keys of the dictionary, we actually route the 50 sine wave oscillators to the first, and only, output of our mixer, and we halve their amplitude. The `Mixer()` class might be a bit difficult to grasp at first, but, as its name implies, it is very efficient when mixing signals.

In line 15 we create a `Sig()` object and we pass the first output of the `Mixer()` as its value. Outputs of the `Mixer()` class are indexed the same way we index lists in Python. In our case we have only one output, but we must still index it like this. We then lower the volume a bit by setting the `mul` attribute to 0.5.

In Figure 3.12 we can see the waveform created by the addition of these sine waves. We can clearly see it is a sawtooth, but it has two differences from the sawtooth we saw at the beginning of this chapter. First of all, it is inverted, going from high to low, whereas the saw-tooth we created with the `Phasor()` class goes from low to high. Second, we can see that throughout its ramp there are small ripples, that are more visible toward the edges. This type of sawtooth is called band-limited, because there is a limit to the number of partials (bands) it contains. A sawtooth created with simpler arithmetic functions, like a rising value from -1 to 1, has no specified limit in its partials. Band-limited waveforms are used to simulate analog oscillators.

The next waveform we will build is the square wave. This is very similar to the sawtooth, with the only difference being that the square wave contains only odd harmonics. The code in Script 3.12 creates a band-limited square wave, and Figure 3.13 illustrates it.

Script 3.12 A band-limited square wave oscillator via Additive Synthesis.

```
1 from pyo import *
2
3 s = Server().boot()
4
5 base_freq = 100
6
7 sines = [Sine(freq=(i*base_freq), mul=(1/i))
8          for i in range(1,80,2)]
9 mixer = Mixer(outs=1, chnls=2)
10
11 for i in range(40):
12     mixer.addInput("additive"+str(i), sines[i])
13     mixer.setAmp("additive"+str(i), 0, 1)
14
15 sig = Sig(mixer[0], mul=.5).out()
16
17 sc = Scope(sig, gain=1)
18 sp = Spectrum(sig)
19
20 s.gui(locals())
```

The main difference between the sawtooth and square wave implementations is in lines 7 and 8, where we create the list with the `Sine()` objects. Since we need only odd harmonics for the square wave, we pass three arguments to the `range()` function, a start value, a stop value, and a step. `range()` returns a list starting from 1, ending at 79, and including only odd numbers, since it increments by two at every step. This list will have 40 values. In line 11 we therefore iterate 40 times to add inputs and set amplitudes to our `mixer`. The rest of the code is the same with the sawtooth version. Figure 3.13 illustrates this band-limited square wave. Again we can see ripples along its path.

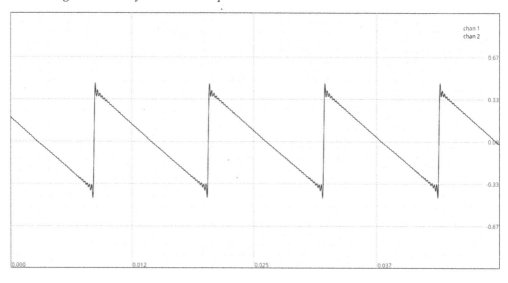

Figure 3.12 A band-limited sawtooth created by adding 50 sine wave oscillators.

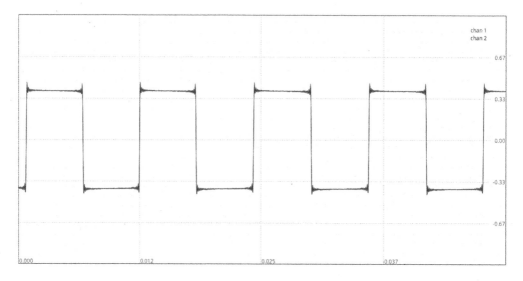

Figure 3.13 A band-limited square wave created by adding 40 sine wave oscillators.

The next waveform will be the triangle. This is slightly more involved than the previous two, because, apart from including only odd partials, the amplitude sign of every other partial is inverted, with the first being positive, the third being negative, the fifth positive, and so on. The code in Script 3.13 creates a band-limited triangle.

Script 3.13 A triangle wave oscillator via Additive Synthesis.

```
1 from pyo import *
2
3 s = Server().boot()
4
5 base_freq = 100
6
7 sines = [Sine(freq=(i*base_freq),mul=(1/pow(i,2)))
8          for i in range(1,80,2)]
9 mixer = Mixer(outs=1, chnls=2)
10
11 for i in range(40):
12     sines[i].mul *= ((((i+1)%2) * 2) - 1)
13     mixer.addInput("additive"+str(i), sines[i])
14     mixer.setAmp("additive"+str(i), 0, .5)
15
16 sig = Sig(mixer[0], mul=.5).out()
17
18 sc = Scope(sig)
19 sp = Spectrum(sig)
20
21 s.gui(locals())
```

Lines 7 and 8 create the list with the sine waves, but the value passed to the `mul` attribute is 1 divided by `i` raised to the second power. This is done with Python's `pow()` function where the first argument is the value to be raised to a power, and the second argument is the exponent. You need to be careful with the number of round brackets you use. For every opening round bracket, there must be a closing one. IDEs with support for various programming languages usually place the closing bracket as soon as you type the opening one.

The next step we need to take is to invert the sign of every other partial. This is done inside the loop that adds inputs and sets amplitudes to the mixer. In line 12 we use the following notation: `*=`. This is equivalent to `+=` which we have used in Chapter 2. This notation multiplies the value to the left of this symbol by the value to the right of it. Its expansion is shown in Script 3.14.

Script 3.14 Expansion of the `*=` semantic.

```
12 sines[i].mul = sines[i].mul * ((((i+1)%2) * 2) - 1)
```

The value to the right results in an alternating 1 and -1. Since our loop will start from 0, we add 1 to it so in this line it starts from 1. We then use the modulo operator to get an alternating 1 and 0 as we have already seen. Multiplying 2 and subtracting 1 will yield a 1 when the result of the modulo is 1, and a -1 when the result of the modulo is 0. This line inverts the sign of the amplitude value of every other partial. The rest of the code is the same as the previous examples. The resulting triangle wave is very similar to the one shown in the beginning of this chapter, even though this is band-limited.

Besides the standard waveforms we can create arbitrary waveforms that don't necessarily follow the same harmonic structure neither for the frequencies nor for the amplitudes. The code in Script 3.15 creates ten sine waves with random frequencies and random amplitudes.

Script 3.15 A band-limited random wave oscillator via Additive Synthesis.

```
 1 from pyo import *
 2 import random
 3
 4 s = Server().boot()
 5
 6 base_freq = 100
 7
 8 sines = [Sine(freq=((random.random()*10)*base_freq),
 9          mul=random.random()) for i in range(10)]
10 mixer = Mixer(outs=1, chnls=1)
11
12 for i in range(10):
13     mixer.addInput("additive"+str(i), sines[i])
14     mixer.setAmp("additive"+str(i), 0, .5)
15
16 sig = Sig(mixer[0], mul=.2).mix(2).out()
17
18
19 def set_rand_freqs():
20     for i in range(10):
21         sines[i].setFreq((random.random()*10)*base_freq)
22
23
24 def set_rand_amps():
25     for i in range(10):
26         sines[i].mul = random.random()
27
28 sc = Scope(sig)
29 sp = Spectrum(sig)
30
31 s.gui(locals())
```

Line 8 utilises the `random()` function of the random module. It might be a bit confusing that both the module and one of its functions have the same name, but you should know that the `random()` function returns a random float between 0 and 1. In this line we multiply the result of `random()` by 10 to scale its range from 0 to 10, and then we multiply this by our base frequency. We use this function to pass a value to the `mul` attribute as well. Note that the resulting waveform will be random, and the more partials we add, the closer it will get to noise. Note also that we set a lower value in the `mul` attribute of the `Sig()` object, as this random configuration can lead to quite loud waveforms.

In lines 19 and 24 we define two functions, one for setting new random frequencies, and one for setting new random amplitudes. If you run this code with Pyo's GUI, you can use the interpreter entry in the Server window to call these functions. Just type the name of the function you want to call (without the `def` keyword) with an empty parenthesis (since none of these take any arguments) and hit the Return key.

Documentation Page:

`Mixer()`: https://belangeo.github.io/pyo/api/classes/pan.html#mixer

3.2.4.1   A More Efficient Additive Synthesis Technique

The Additive Synthesis examples we have seen so far are intended for explaining how the various waveforms are created, but they are not so efficient as they need a lot or resources, since for a single waveform we need 40 to 50 oscillators. A more efficient way to create waveforms with Additive Synthesis is to make all these calculations and store the desired waveform in a table. This table will then contain one period of the waveform of our oscillator and we can simply read its data at the desired rate to get back the pitch we want. This type of oscillator is called a table-lookup oscillator. Pyo provides tables for the three waveforms, triangle, sawtooth, and square. The Sine() class is a table-lookup oscillator itself, so there is no need to provide a class to create this waveform. The code in Script 3.16 creates a triangle wave oscillator using Pyo's TriangleTable() class. To create a sawtooth and a square wave band-limited oscillator, use the SawTable() and SquareTable() classes instead, respectively.

Script 3.16 A band-limited table-lookup triangle wave oscillator.

```
 1 from pyo import *
 2
 3 s = Server().boot()
 4
 5 tritab = TriangleTable(order=40).normalize()
 6 lookup = Osc(table=tritab, interp=2, freq=200, mul=.2).out()
 7
 8 sc = Scope(lookup, gain=1)
 9
10 s.gui(locals())
```

The order kwarg is the number of partials we want our waveform to include. With this approach, we can use as many partials as we want, and the only strain on the CPU will be at the creation of the table. Once the waveform is stored in the table, the rest of the code is very lightweight. One feature these Additive Synthesis tables include is the normalize() method. This is necessary to be called as, without it, the resulting waveform will exceed the digital audio limits, and will go below -1 and above 1. normalize() will bring the waveform back to the -1 to 1 range. TriangleTable() has one more kwarg: size. This sets the size of the table, and defaults to 8192 (2 to the 13th power).

In line 6 we create an Osc() object. This is Pyo's table-lookup oscillator. It takes the same kwargs as the Sine() class, plus a table and an interp kwarg. The table kwarg sets the table this oscillator will look up to read a waveform. In this case it is the tritab table object. The interp attribute sets the interpolation method, in case the reading index of the oscillator falls in between points. The resulting waveform is stored in a table with the size set to it via the size argument (in our case it is 8192, the default). A table-lookup oscillator increments an index from 0 up to the size of the table -1, and outputs the value stored at each index. In computer programming, tables are indexed with integers, but the reading index of a table-lookup oscillator is very likely to fall in between two indexes, depending on the frequency we set.

If we set no interpolation and the reading index falls in between two points, the decimal part of the index will just be truncated, and its integral part will be used to read from the table. If we use linear interpolation (which is the default), then the oscillator will output a value that is half way between the values at the two integer indexes around its reading index. For example, if the reading index of the table-lookup oscillator is 1000.5, and the table that contains the waveform has stored the value 0.44 at index 1000 and the value 0.46 at index 1001, then the oscillator will output 0.45. The other two interpolation methods available are cosinus and cubic. These are more involved concerning the maths, and more CPU expensive, but they do provide a more sophisticated interpolation. For our current needs, the linear interpolation is enough. The interpolation methods are indexed as follows: 1 for no interpolation, 2 for linear, 3 for cosinus, and 4 for cubic.

Another nice feature when using this kind of Additive Synthesis is that we don't need to set a base frequency when we create our waveforms. The frequency is set in the table-lookup oscillator and not the table itself. Apart from these three waveforms, we can create any waveform we like by using Pyo's `HarmTable()` class. The code in Script 3.17 creates a waveform with ten partials and random amplitudes.

Script 3.17 A band-limited table-lookup random wave oscillator.

```
 1 from pyo import *
 2 import random
 3
 4 s = Server().boot()
 5
 6 amps = []
 7
 8 for i in range(10):
 9     amps.append(0)
10
11 randtab = HarmTable(amps)
12 randtab.autoNormalize(True)
13 lookup = Osc(table=randtab, freq=200, mul=.2).out()
14
15 def rand_amps():
16     for i in range(10):
17         amps[i] = random.random()
18     randtab.replace(amps)
19
20 rand_amps()
21
22 sc = Scope(lookup, gain=1)
23
24 s.gui(locals())
```

In line 6 we create an empty list, and in line 8 we run a loop that fills this list with ten zeros. In line 11 we create a `HarmTable()` object and pass our list as its argument. This list contains the amplitudes of the partials. As with `TriangleTable()`, `HarmTable()` takes one more argument, `size`, which sets the size of the table that will be created, again defaulting to 8192. In line 12 we call the `autoNormalize()` method of the `HarmTable()` class and set it to `True`, so that our waveform will be normalized.

In line 15 we define a function that sets random values to our amplitudes list, and when this is done, we replace the list of our `HarmTable()` object by calling its `replace()` method. We pass the amps list again as the argument, since the list has now changed. In line 20 we call this function to initialise our table with a waveform. If you run this code and call the `rand_amps()` function from the interpreter, you will see the waveform change every time. Note that this sound is harmonic as we don't have access to the frequency ratios, which are integer multiples of the base frequency. If you want to create an inharmonic waveform, you will have to use the method we used earlier where we used separate oscillators.

Documentation Pages:

TriangleTable(): https://belangeo.github.io/pyo/api/classes/tables.html#triangletable
SawTable(): https://belangeo.github.io/pyo/api/classes/tables.html#sawtable
SquareTable(): https://belangeo.github.io/pyo/api/classes/tables.html#squaretable
Osc(): https://belangeo.github.io/pyo/api/classes/tableprocess.html#osc
HarmTable(): https://belangeo.github.io/pyo/api/classes/tables.html#harmtable

3.2.5   *Granular Synthesis*

Granular Synthesis is a technique used by many electronic music composers. Iannis Xenakis was a pioneer in Granular Synthesis with compositions applying this concept dating back to 1958 (Xenakis, 1992, p. 79). The idea of this technique is to organise very short sound segments in various ways, either to transform sound in pitch or frequency, or to create completely new sounds. These short segments are derived from a stored waveform which can be a synthesized waveform, like the ones we saw in the Additive Synthesis section, or recorded sounds. Typically, composers apply Granular Synthesis on recorded sounds, but using this technique with more standard waveforms can also provide interesting results.

Ingredients:

- Python3
- A text editor (preferably an IDE with Python support)
- The Pyo module
- the wxPython module, if you want to display waveforms
- A terminal window in case your editor does not launch Python scripts

Process:

Pyo includes two Granular Synthesis generators in its classes, Granulator() and Granule(). These are similar with a few differences. In Granular Synthesis there are a few parameters that are settable by the user. These are typically the number of grains or their density, the pitch, the duration of the grains, the envelope applied to the grains (more on envelopes later on in this chapter), and the position of the index, also referred to as a pointer, in the waveform table.
    Granulator() can explicitly set the number of grains, whereas Granule() sets the density of grains per second. Granulator() also includes a basedur attribute that controls the speed of the pointer to read the grains at their original pitch. Granule() can change between two granulation modes, synchronous and asynchronous. In synchronous mode, the grains are being triggered at regular intervals, while in asynchronous mode, grains are being triggered more sparsely, creating a sound that is more spread. The default mode is synchronous. The code in Script 3.18 is a combination of the examples from the documentation of Pyo, and it includes both Granular Synthesis generators.

Script 3.18 Granular Synthesis with Pyo's Granulator() and Granule() classes.

```
1 from pyo import *
2
3 s = Server().boot()
4
```

```
 5 snd = SndTable(SNDS_PATH+"/transparent.aif")
 6 env = HannTable()
 7 pos = Randi(min=0, max=1, freq=[0.25, .3], mul=snd.getSize())
 8 dur = Noise(.001, .1)
 9
10 grn1 = Granulator(snd, env, pitch=[1, 1.001], pos=pos,
11                     dur=dur, grains=24, mul=.1)
12 grn2 = Granule(snd, env, pitch=[1, 1.001], pos=pos, dur=dur,
13                   dens=24, mul=.1)
14
15 selvoice = SigTo(0, time=5)
16 sel = Selector(inputs=[grn1, grn2], voice=selvoice).out()
17
18 grn1.ctrl()
19 grn2.ctrl()
20
21 sc = Scope(sel)
22
23 s.gui(locals())
```

We have inserted a few new elements in our code that need explanation. Line 5 creates a sound table with a sample that comes with Pyo. The SDNS_PATH is a string containing the path to the directory where Pyo stores some sound samples. This is set when Pyo is installed in your system. One of these sound files is "transparent.aif", which we use in this example. Line 6 creates a Hanning table that is used as the envelope for the grains. This is essentially a raised unipolar cosine going from 0 to 1 and back to 0, and it is used as a window within which a grain sounds. Since this envelope starts from 0, when a grain is triggered, it will start with 0 amplitude. It will then rise to 1 in a cosine shape, and then go back to 0. This way clicks are avoided and the resulting sound is smooth.

Line 7 creates a random value generator. Randi() creates pseudo-random float values with linear interpolation between them. Its kwargs should be self-explanatory. Note that we pass a list to the freq kwarg. As we have already seen, this will create a list of Randi() objects which will share all kwargs except from the ones that are passed as lists, in which case each object will get the corresponding kwarg in this list. The mul attribute gets the size of the sound table object, which is the size in samples of the recorded sound we load in this table.

Line 8 creates a noise generator. The Noise() class takes only two kwargs, here provided as positional arguments, mul and add. Lines 10 and 12 create two objects, one for each Granular Synthesis generator class of Pyo. The first two arguments of these classes are positional, and these are the table that contains the samples, and the envelope to be applied to the grains. The rest are kwargs. In this example we pass quite a lot of signal generators to the kwargs of the Granular Synthesis generators. From now on, we will refer to such signal generators as PyoObjects, PyoTableObjects, or a similar name. Passing PyoObjects to kwargs will animate the sound since all these parameters will have their value constantly change.

Line 15 creates a smoothing signal which can take a value in its first argument to go to from its current one in the time specified by the time kwarg. This is set in seconds, and, in our example, we ramp to this value in 5 seconds. Line 16 creates an object that can take an arbitrary number of inputs and interpolate between adjacent inputs to give a single output. The inputs are provided as a list and, in this example, we pass the two Granular Synthesis generators. The voice kwarg sets the interpolation value between the inputs, starting from 0 for the first input, 1 for the second, etc. This is the signal we output to the speakers and we also send to the `Scope()`. Lines 18 and 19 call the `ctrl()` method of both Granular Synthesis generators.

If you run this code you will get the Pyo Server window, the oscilloscope window, and two windows with control widgets, one for each Granular Synthesis generator. Note that the parameters that take PyoObjects do not appear in the widgets. If you want to control these parameters manually, you will have to provide a constant value to their respective kwarg, like we do with the parameters that appear in the widgets. The resulting sound is stereo even though we did not call the `mix()` method, since some of the parameters of the Granular Synthesis generators, or some parameters of the PyoObjects passed to kwargs of the Granular Synthesis generators (e.g. the `freq` kwarg of the `Randi()` class), are provided as lists. If these lists contain two items, then we will get two audio streams out of the objects that create sound. By calling their `out()` method without arguments, the two sound streams will be diffused to our stereo system.

In the Pyo Server window you can use the interpreter entry to change the value of the `selvoice` object, by typing the following line:

```
selvoice.setValue(1)
```

If you type this you will hear a shift from one Granular Synthesis generator to the other. This shift will take 5 seconds, which is the time we have set to the `selvoice` object. Since our `Selector()` has only two inputs, the value we can pass to `selvoice` can be between 0 and 1, but it can be a float somewhere in between these two values. If you want to animate this process, change line 15 of the code in Script 3.18 with the line in Script 3.19.

Script 3.19 Animating the interpolation between the two Granular Synthesis generators.
```
15 selvoice = Sine(freq=.1, mul=.5, add=.5)
```

Play with the paremeters that are controllable with the widgets to change the overall sound. You can also change any attribute of any object in the code using the interpreter entry of the Server window. For example, if you want to change the `mul` attribute of the `pos` object, type the following line in the interpreter entry:

```
pos.mul = 1500
```

Since the `Granule()` class is capable of switching between synchronous and asynchronous granulation mode, you can also try to change that. In the interpreter entry, type the following line:

```
grn2.setSync(False)
```

To switch back to synchronous mode, call the same method with a `True` argument. You can also try a different table. A `SawTable()` can be an option. If you want to try this, change line 5, and instead of an `SndTable()` change it to a `SawTable()` with the corresponding arguments.

Documentation Pages:

`SndTable()`: https://belangeo.github.io/pyo/api/classes/tables.html#sndtable
`HannTable()`: https://belangeo.github.io/pyo/api/classes/tables.html#hanntable
`Randi()`: https://belangeo.github.io/pyo/api/classes/randoms.html#randi

Noise(): https://belangeo.github.io/pyo/api/classes/generators.html#noise
Granulator(): https://belangeo.github.io/pyo/api/classes/tableprocess.html#granulator
Granule(): https://belangeo.github.io/pyo/api/classes/tableprocess.html#granule
SigTo(): https://belangeo.github.io/pyo/api/classes/controls.html#sigto

3.3   Effects

Our next topic will be effects. Pyo includes various classes for different types of effects like distortion, delay, reverb, chorus, and others. In this section we will cover some of them and provide sources for documentation of all the available effects classes. The first effect we will cover is filtering, and together with it, we will cover another audio synthesis technique, Subtractive Synthesis. Although this could fit in the previous section with the rest of the synthesis techniques, Subtractive Synthesis is tightly connected with filters, so it will be covered here.

3.3.1   *Subtractive Synthesis and Filtering*

Subtractive Synthesis is a technique where we generate a waveform with a rich spectrum and then we filter it to get a different waveform (Huovilainen and Välimäki, 2005, p. 300). A discussion on filters is necessary before we start writing code that applies Subtractive Synthesis. According to Rabiner *et al.*, "a digital filter is a computation process or algorithm by which a digital signal or sequence of numbers (acting as input) is transformed into a second sequence of numbers termed the output digital signal" (Rabiner *et al.*, 1972, p. 326). That means that any signal processing is an act of filtering. Nevertheless, when we speak of filters we refer to the action of attenuating or boosting certain frequencies of the spectrum of a sound (Roads, 1996, p. 185).

We can roughly separate filters into two categories, pass and shelving. The former lets a certain portion of the spectrum pass while it attenuates the rest, and the latter boosts or attenuates a certain part of the spectrum while it leaves the rest intact. Typical pass filters are the lowpass, the bandpass, and the highpass. As their names imply, the lowpass filter lets frequencies below a certain threshold pass, the bandpass lets a band of frequencies around a certain threshold pass, and the highpass lets frequencies above a certain threshold pass. This threshold is called the cutoff frequency. Figure 3.14 illustrates the spectrum of white noise passed through a lowpass filter with a cutoff frequency set to 3000Hz.

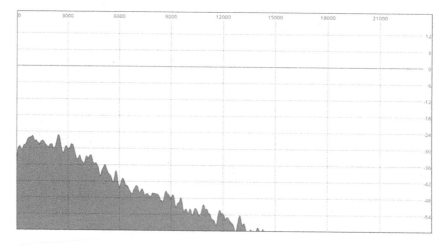

Figure 3.14 Spectrum of white noise passed through a lowpass filter with a 3000Hz cutoff frequency.

White noise is a signal that contains all the audible frequencies. When unaltered, its spectrum should spread equally across the audible range. In Figure 3.14 we see a noisy spectrum that starts attenuating at 3000Hz, and goes completely silent after 12000Hz. The range of attenuation of the spectrum is called the slope of the filter. The filter applied to this signal is a second-order lowpass filter. The higher the order of the filter, the sharper its slope. A four-order lowpass filter would result in the slope of the spectrum to drop to silence somewhere around 6000Hz.

If we pass white noise through a highpass filter, the spectrum will be the opposite of Figure 3.14. It will start from silence at the left part of the spectrum, and its amplitude will rise as the frequencies go higher, toward the right part of the spectrum. As with the slope of the lowpass, the same applies to the slope of the highpass, with higher-order filters having a sharper slope. The bandpass filter is somewhat different as it lets a band of frequencies around its cutoff frequency pass. In this filter we have control over the cutoff but also over the width of the band. The latter is called the Q, an abbreviated term standing for "Quality Factor". Figure 3.15 illustrates white noise passed through a second-order bandpass filter with a cutoff frequency of 12000Hz and a Q factor of 15. In bandpass filters, the higher the Q, the narrower the band, and vice versa. Similarly to the bandpass, there is the band-reject filter. This is the opposite of the bandpass, as it cuts off a band of frequencies around its cutoff. The width of the band is again controlled by the Q factor. The order of the filter affects its slope, as with the lowpass and highpass.

The shelving filters are different in that they alter the frequencies of their region instead. A lowshelf filter alters the frequencies below its cutoff frequency, in contrast to the lowpass which attenuates frequencies above its cutoff. Shelving filters can both attenuate and boost frequencies in their region. Figure 3.16 illustrates white noise passed through a lowshelf filter that boosts frequencies below 8000Hz.

A third filter type is the resonant filter. This filter is similar to the pass filters with the addition of adding resonance to the cutoff frequency. A typical resonant filter is the Moog filter, named after the synthesizer pioneer, Robert Moog. Pyo's implementation of this filter is the `MoogLP()` class, standing for Moog Low Pass.

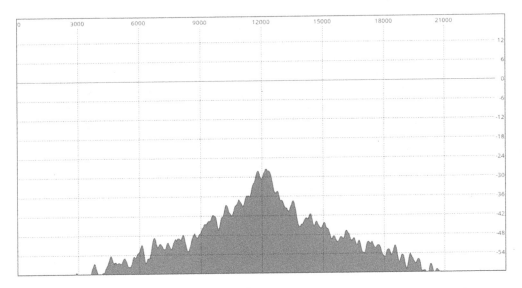

Figure 3.15 Spectrum of white noise passed through a bandpass filter with a 12000Hz cutoff frequency and a Q factor of 15.

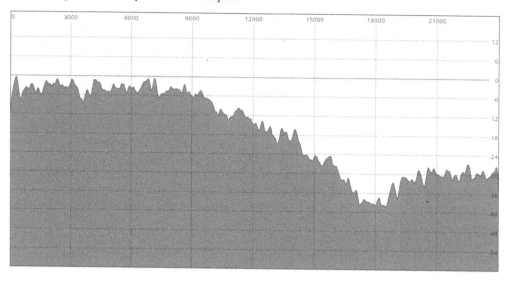

Figure 3.16 Spectrum of white noise passed through a lowshelf filter with a 8000Hz cutoff frequency.

Ingredients:

Python3
- A text editor (preferably an IDE with Python support)
- The Pyo module
- the wxPython module, if you want to display waveforms
- A terminal window in case your editor does not launch Python scripts

Process:

The MoogLP() filter is a fourth-order lowpass filter. We will use it in our Subtractive Synthesis example, where we will filter a square wave oscillator to obtain a waveform with less harmonics. The code is shown in Script 3.20.

Script 3.20 Subtractive Synthesis with a square wave oscillator passed through a resonant lowpass filter.

```
1 from pyo import *
2
3 s = Server().boot()
4
5 freq = Sig(80)
6
7 saw = Phasor(freq, mul=2, add=-1)
8 sqr = Round(saw)
9 filt = MoogLP(sqr, freq=freq*5, res=1.25)
10
11 outsig = Mix(filt.mix(1), voices=2, mul=.2).out()
12
13 filt.ctrl(map_list=[SLMap(0, 2, 'lin', 'res', 1.25)])
```

```
14
15 sc = Scope(outsig, gain=1)
16 sp = Spectrum(outsig)
17 s.gui(locals())
```

In lines 7 and 8 we create a square wave oscillaor, like we did in Script 3.2 and we pass that through the `MoogLP()` filter. Besides the `freq` attribute, `MoogLP()` has a `res` attribute, which sets the resonance of the filter. According to Pyo's documentation, when this value is set to 0 there is no resonance, and when set to 1 there is medium resonance. When this value is higher than 1, self-oscillation occurs. Since we call the `ctrl()` method of our filter object, we cannot use a value higher than 1. Even if we hard-code such a value in line 9, as soon as the GUI launches, the value will go down to 1, as this is the maximum value set to this attribute by default. To remedy this, Pyo provides the `SLMap()` class, a list of which we can pass as an argument to a `ctrl()` method. Its positional arguments are the minimum and maximum value, its scale – linear or logarithmic, passed with the strings "`lin`" and "`log`" respectively – the name of the attribute the values of `SLMap()` will be sent to, and an initial value. In this case we control the `res` attribute of the filter, so we can use Pyo's GUI and still be able to set this value above 1.

The cutoff frequency passed to the filter is five times the frequency of our oscillator. When we set a high value to the resonance attribute of the filter, the fifth harmonic of the oscillator will be boosted, giving a different characteristic to our sound. You can change the base frequency by calling `freq.setValue(new_value)` in the interpreter entry of Pyo's server window, where `new_value` will be the frequency we will set to our oscillator. The cutoff frequency of our filter will always be five times this value.

One last note before we end this section is about the order of filters. We have mentioned second and fourth-order filters, and how this affects their slope. Pyo's classes include a large array of different types of filters, for various applications. Most of these filters are second-order, but we can create higher-order filters by stacking lower-order filters together. This means that we can pass a signal through a second-order filter, and then pass this filter to another second-order filter. This will result in a fourth-order filter. Even though this practice is common, Pyo includes classes that provide this filter stacking internally, to save on CPU and memory, and it is better to use those, if you need higher-order filtering. Check the documentation pages of these classes for information.

Documentation Pages:

`MoogLP()`: https://belangeo.github.io/pyo/api/classes/filters.html#mooglp
`SLMap()`: https://belangeo.github.io/pyo/api/classes/map.html#slmap
Pyo's filter classes: https://belangeo.github.io/pyo/api/classes/filters.html

3.3.2 *Delay*

The delay is a celebrated effect in electronic music. In fact, most sound processes use delays, including filters, chorus, and reverberation (Farnell, 2010, p. 214). The delay though can be simply used as the actual effect, something that is very common.

Ingredients:

- Python3
- A text editor (preferably an IDE with Python support)

- The Pyo module
- the wxPython module, if you want to use Pyo's GUI
- A terminal window in case your editor does not launch Python scripts

Process:

To utilise a delay line, we will need some audio input. Usually, sounds with distinctive attacks work better with delay lines, because the attack is then audible within the delayed signal, and it can even be sustained and altered, if we feed the output of the delay line back to its input. Script 3.21 is copied from Pyo's documentation.

Script 3.21 Delaying an audio sample.
```
1 from pyo import *
2
3 s = Server().boot()
4
5 sample = SfPlayer(SNDS_PATH + "/transparent.aif", loop=True,
6                   mul=.3).mix(2).out()
7 delay = Delay(sample,delay=[.15,.2],feedback=.5,mul=.4).out()
8
9 s.gui(locals())
```

In this example we use the SfPlayer() class, which is a sound file player. The kwargs passed in this example are the path to the sound file you want to play, and the loop state. In Script 3.21 we can hear both the original signal, and its delayed copy. Since the audio sample is not stereo, to get the original sound in stereo, we use the mix(2) attribute to convert the sample object to a Mix() object. In the Delay() class we pass a list as the argument to the delay attribute. This will create a stereo stream of the Delay(.) object, whether the input is stereo or not. In this example, the delay times – expressed in seconds – are not very short, and the feedback amount is set to 0.5. This means that the output of the delay line will be halved in amplitude before it is sent back to its input.

Another nice way to use a delay line is with natural feedback, combined with high internal feedback values. Script 3.22 is an example of sending the audio captured by a microphone to a delay line with a short delay time and a high feedback. If the microphone is close to the speakers, then natural feedback will occur. This can be done by using the internal speakers and microphone of a laptop, or by placing the microphone close to the speakers. You have to be careful though, since, if the microphone is too close, or the speakers are too loud, then the natural feedback will grow in amplitude exponentially, and it is possible to damage your equipment, or your ears.

Script 3.22 Delaying the audio captured by a microphone.
```
1 from pyo import *
2
3 s = Server().boot()
4
5 mic = Input()
6 deltime = Sine(freq=[.1, .12], mul=.015, add=.04)
7 feedback = Sine(freq=[.09, .11], mul=.045, add=.855)
8 delay=Delay(mic,delay=deltime,feedback=feedback,mul=.2).out()
9
10 s.gui(locals())
```

In Script 3.22 we get the audio captured by a microphone through the `Input()` class. The two sine wave oscillators that control the delay time and the feedback are created with lists in their kwargs, so a stereo stream is created for each and then passed to the `Delay()` object, so we don't need to call the `mix(2)` method. If you run this example, make sure to have your mouse over the amplitude slider in Pyo's Server window, and play with it to create, and most importantly, to control the amount of natural feedback.

Documentation Pages:

`SfPlayer()`: https://belangeo.github.io/pyo/api/classes/players.html#sfplayer
`Input()`: https://belangeo.github.io/pyo/api/classes/generators.html#input
`Delay()`: https://belangeo.github.io/pyo/api/classes/effects.html#delay

3.3.3   Reverb

Another common effect in electronic music is the reverb. This is a natural effect that occurs when sounds reflect on surfaces. Reverberation is more apparent in closed and large spaces, where there are many surfaces of different size, shape, and angle, compared to the sound source. As sound propagates in space, it arrives at the listener directly, given that there are no physical obstacles between the listener and the sound source, but it also reflects off the surfaces of the surrounding space. These reflections arrive at the listener with a short delay, compared to the direct propagation. Additionally, the reflected sound usually is filtered, as air and many surfaces tend to filter high harmonics (Roads, 1996, p. 472). This combination of the dry sound with the delayed and filtered reflections, results in what we perceive as reverberation.

Ingredients:

- Python3
- A text editor (preferably an IDE with Python support)
- The Pyo module
- the wxPython module, if you want to use Pyo's GUI
- A terminal window in case your editor does not launch Python scripts

Process:

The reverb effect can be achieved by combining delay lines with filters, but that would require a lot of fine tuning, and it would also be CPU expensive. Pyo provides two reverberators that are intuitive, so we will use those. The code in Script 3.23 utilises the `Freeverb()` class, while the code in Script 3.24 utilises the `WGVerb()` class. Even though digital reverbs simulate natural reverberation, we can use them in a more unnatural way. Both Scripts 3.23 and 3.24 include a function that clears the memory of the reverberators, sets the signal to 100% dry instantly, and then restores the reverberation. This results in an unnatural cut of the reverb, as if the room that the reverb simulates shrinks in an instance, and then grows big again.

Script 3.23 Using the Freeverb() class.

```
1 from pyo import *
2
3 s = Server().boot()
4
```

```
 5 bal = SigTo(.75)
 6
 7 samp = SfPlayer(SNDS_PATH+"/transparent.aif",loop=True,mul=.4)
 8 verb = Freeverb(samp, size=[.79,.8], damp=.9, bal=bal).out()
 9
10 def restore_verb():
11     bal.setTime(5)
12     bal.setValue(.75)
13
14 ca = CallAfter(restore_verb, time=1)
15 ca.stop()
16
17 def reset_verb():
18     verb.reset()
19     bal.setTime(0)
20     bal.setValue(0)
21     ca.play()
22
23 s.gui(locals())
```

The two reverberators provided by Pyo have a different approach to creating this effect. Freeverb() is an implementation of "Jezar at Dreampoint"'s Freeverb, a public domain C++ reverb program (Smith, 2010). The size attribute is the length of the reverb, simulating the size of the room. The damp attribute controls the attenuation of high frequencies. The higher this value, the faster the high frequencies will be attenuated. The bal attribute controls the balance between the dry and wet signals. The dry signal has no reverberation, so a value of 0 for the bal attribute will result in a sound without reverb, while a value of 1 will result in the reverberated sound only.

In line 5 of Script 3.23 we create a SigTo() object which is very similar to Sig() with the addition of making a linear interpolation between the previous and the current value, the duration of which is set via the time attribute, defaulting to 0.025 seconds. We set its value to 0.75 and we pass this as the value of the bal attribute of the reverberator. In line 17 we define a function that clears the memory of the reverberator by calling its reset() method. Then we set both the time and the value of the SigTo() object to 0, so we will immediately get the dry sound only.

A few lines before the reset_verb() function, we create a CallAfter() object, in line 14. The CallAfter() class takes a function name as its first argument. It will then call this function after the time set to its time attribute has passed, since the creation of the CallAfter() object. To avoid calling its function as soon as we launch our script, we call its stop() method immediately after, in line 15. This object is created after the restore_verb() function, because this is the function we pass to its first argument. If we create this object before the function, we will get a NameError, since at the initialisation of it, its calling function will have not yet been defined. Using a CallAfter() object enables us to delay calling a function without stalling our program, which would happen if we used time.sleep(1) instead. For the purposes of this script, using time.sleep(1) would have the same result, but now it is a good point to acquaint ourselves with this Pyo class, as it will be useful in later projects of this book.

At the end of the reset_verb() function, in line 21, we call the play() method of the CallAfter() object, to start counting and call its function after its time has passed. This will result in `CallAfter()` calling the `restore_verb()` function exactly one second after we call its `play()` method. This function is defined in line 10, and all it does is set the time of the `SigTo()` object to 5 seconds, and its value to 0.75. Once this function is called, the `bal` attribute of our reverberator will ramp from 0 to 0.75 in five seconds.

Script 3.24 if very similar to Script 3.23, with the exception of the reverberator used. Here we use the `WGVerb()` class, which stands for Wave Guide reVerb. The arguments of this class are different than the arguments of `Freeverb()`. `WGVerb()` is based on Feedback Delay Networks (FDN) (Smith, 2010) and filters, and its arguments specify the feedback of the delay lines, the cutoff frequency of the filters, and the balance between the dry and wet signals.

In this script, we define the same reset and restore functions as in Script 3.23. When you run any of these two scripts, in the interpreter entry of Pyo's Server window, you can call the `reset_verb()` function and hear the result. The sound will immediately switch to dry, and after one second it will ramp back to its previous setting for five seconds.

Although this section covered three effects only, there are many more available through Pyo's classes including distortion, frequency shifting, chorus, and others. You can check the documentation pages of each of these classes. We have already covered a lot of ground, so Pyo's documentation should be quite clear so you can try out things on your own.

Script 3.24 Using the WGVerb() class.

```
 1 from pyo import *
 2
 3 s = Server().boot()
 4
 5 bal = SigTo(.6)
 6
 7 samp = SfPlayer(SNDS_PATH+"/transparent.aif",loop=True,mul=.4)
 8 verb = WGVerb(samp, feedback=[.74,.75], cutoff=5000,
 9               bal=bal,mul=.3).out()
10
11 def restore_verb():
12     bal.setTime(5)
13     bal.setValue(.6)
14
15 ca = CallAfter(restore_verb, time=1)
16 ca.stop()
17
18 def reset_verb():
19     verb.reset()
20     bal.setTime(0)
21     bal.setValue(0)
22     ca.play()
23
24 s.gui(locals())
```

Documentation Pages:

`Freeverb()`: https://belangeo.github.io/pyo/api/classes/effects.html#freeverb
`WGVerb()`: https://belangeo.github.io/pyo/api/classes/effects.html#wgverb
`CallAfter()`: https://belangeo.github.io/pyo/api/classes/pattern.html#callafter
Pyo's effects classes: https://belangeo.github.io/pyo/api/classes/effects.html

3.4   Control

Now that we have learned how to create sound and how to apply certain effects to it, it is time to learn how to control our sounds over time, to make more musical patterns. Pyo includes classes for both control signals and events sequencing. In this section we will combine classes from these two categories to control our sounds.

3.4.1   *Applying Envelopes*

Most of the sounds we have created so far are produced by oscillators. We have focused on the timbre of the sounds we create, but we have not yet dealt with controlling the lifetime of a sound. Of course we can control the amplitude of a sound manually, by setting various values to the `mul` attribute of an oscillator, but it becomes more interesting when we can control the amplitude programmatically. This is done with what is called an envelope. This is the "primary tool for shaping the loudness and other qualities of a note or other sound event" (Bjørn and Meyer, 2018, p. 220), not only in digital audio, but in analog too.

A typical envelope is shown in Figure 3.17. This envelope is the most common in electronic music, and it is called an ADSR envelope, standing for Attack-Decay-Sustain-Release, the four stages of the envelope. The black outline is the actual envelope shape, and the grey vertical lines are indicating the separation between the four stages of the envelope. The X-axis in Figure 3.17 represents time, and the Y-axis represents control values. Typically, an ADSR is used to control the amplitude of an oscillator, so in this case, the Y-axis would represent amplitude values. Its shape is a crude simulation of the evolution of sound in acoustic instruments, where the sound takes a short amount of time to rise to its full amplitude, the Attack part, then drops a bit lower in volume, the Decay part, keeps this lower volume for some time, the Sustain part, and then drops to silence, the Release part.

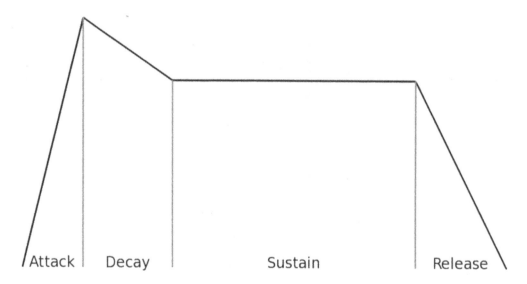

Figure 3.17 An ADSR envelope with its four stages separated by the grey vertical lines.

Ingredients:

- Python3
- A text editor (preferably an IDE with Python support)
- The Pyo module
- the wxPython module, if you want to use Pyo's GUI
- A terminal window in case your editor does not launch Python scripts

Process:

Pyo includes two classes that generate envelopes, `Adsr()` and `Fader()`. The former creates an ADSR envelope, as its name implies, and the latter creates a simpler, AD (Attack-Decay) envelope. This is a two-stage envelope that rises and falls, without a sustain part. Both envelopes can have a pre-defined duration, or they can hold a stage until their `stop()` method is called. Script 3.25 makes use of an ADSR envelope on an FM synth.

Script 3.25 Applying an ADSR envelope to the amplitude and index of an FM synth.

```
1 from pyo import *
2
3 s = Server().boot()
4
5 adsr = Adsr(attack=.05,decay=.2,sustain=.6,release=.1,dur=2)
6 fm = FM(carrier=200, ratio=.248, index=adsr*8,
7         mul=adsr*0.3).mix(2).out()
8
9 s.gui(locals())
```

In line 5 we set all the parameters of our envelope. All kwargs but the `sustain` represent time in seconds which sets their duration, while the `sustain` kwarg represents amplitude. In the initialisation of the `FM()` object, we pass the `Adsr()` object to two of its attributes, index and `mul`. With the `mul` attribute we control the amplitude of the synth, and with the `index` we animate the index of the Frequency Modulation. Remember that the index in FM controls the number of sidebands. The higher the index, the more the sidebands, and vice versa. The more sidebands, the more brilliant the sound is. Note that we can do simple arithmetic operations on PyoObjects. In this case, we multiply the envelope's output by 8 for the index and by 0.3 for the amplitude, to soften the sound.

  Run this script and in the interpreter entry type `adsr.play()` to activate the envelope. You will hear sound lasting two seconds, the timbre of which evolves over time, since we control the index with the envelope as well. The resulting sound is similar to the sound of a digeridoo, with the attack bringing out higher harmonics, that fade out shortly after. If you want the envelope to hold its sustain stage until you call its `stop()` method, set its duration to 0 by typing `adsr.dur=0`.

Documentation Pages:

`Adsr()`: https://belangeo.github.io/pyo/api/classes/controls.html#adsr
`Fader()`: https://belangeo.github.io/pyo/api/classes/controls.html#fader

3.4.2   Sequencing

The next step in controlling our sounds is to create a sequence of events instead of activating and deactivating our envelopes manually. Similarly to other audio programming languages like SuperCollider (Wilson, Cottle and Collins, 2011, p. 179) and Csound (Lazzarini *et al.*, 2016), Pyo includes classes for event triggering and score creation, as well as for sample accurate timing. By score, we refer to a series of events, rather than a traditional Western-music notation score. The three classes for event sequencing in Pyo are `CallAfter()` that we have already seen, `Pattern()`, and `Score()`. Part of the sample accurate timing classes are `Metro()` and `TrigEnv()`. Pyo includes an entire Events Framework too, but this framework is large to the extent that it needs its own chapter. It is discussed and analysed in Chapter 10.

3.4.2.1   Pattern()

The first event sequencing class we will examine is `Pattern()`. This class takes a function name as its first argument, and a duration in seconds. When it is activated, it calls its function at regular intervals, based on the duration we have set.

Ingredients:

- Python3
- A text editor (preferably an IDE with Python support)
- The Pyo module
- the wxPython module, if you want to use Pyo's GUI
- A terminal window in case your editor does not launch Python scripts

Process:

The example we will look at will build on top of the envelope example. The code in Script 3.26 adds some code to that of Script 3.25. We have changed the duration of the envelope to 0, so we can call its `stop()` method to stop it. We have also shortened the overall duration of the envelope by setting the decay to 0.1.

Script 3.26 Controlling an ADSR envelope with the Pattern() class.

```
 1 from pyo import *
 2
 3 s = Server().boot()
 4
 5 adsr = Adsr(attack=.05,decay=.1,sustain=.6,release=.1,dur=0)
 6 fm = FM(carrier=200, ratio=.248, index=adsr*8,
 7         mul=adsr*0.3).mix(2).out()
 8
 9 counter = 0
10 def playenv():
11     global counter
12     if not counter:
13         adsr.play()
14     else:
15         adsr.stop()
```

```
16      counter += 1
17      if counter == 2:
18          counter = 0
19
20 pat = Pattern(playenv, time=.2).play()
21
22 s.gui(locals())
```

In line 9 we create a variable that will function as a counter to determine whether we will call the `play()` or the `stop()` method of the envelope. This variable is used inside the `playenv()` function. In this function we must declare this variable as global, otherwise we will get an `UnboundLocalError` stating that this variable is referenced before its assignment. In the `playenv()` function we test the value of the `counter`. If it is 0, then the test in line 12 will be successful, as it tests against the opposite of the value of `counter` by using the keyword `not`. This is a Python reserved keyword that outputs a Boolean depending on the value it is used against. If this value is 0, then `not` will output `True`, whereas if the value is not 0, it will output `False`. Remember from chapter 2 that the `True` and `False` Booleans are treated like numbers. Any number that is not 0 is regarded as `True` and 0 is regarded as `False`.

By using this conditional test in lines 12 and 14, we will either trigger our envelope, or we will stop it. After the test we increment our counter by 1 and reset it to 0 when it reaches 2. If we omit this last test in line 17, our envelope will be triggered only once, as our counter will never reset to 0 and `not` will keep on outputting `False`.

In line 20 we initialise our `Pattern()` object and pass our function to its first argument. We also set the time attribute to 0.2, and we trigger the object by calling its `play()` method. Every 0.2 seconds the `playenv()` function will be called, and the envelope will alternate between being triggered and stopped.

You might have noticed that many PyoObjects have a `play()` and a `stop()` method. You might have also noticed that these methods are not always included in the documentation pages of Pyo's classes. This is because these methods are part of what is called a super class in Python, which all Pyo classes inherit from. These concepts concern classes and we will examine them further in Chapter 8.

Documentation Page:

`Pattern()`:https://belangeo.github.io/pyo/api/classes/pattern.html#pattern

3.4.2.2 Score()

The second event sequencing class we will examine is `Score()`. This is somewhat similar to `Pattern()`, but it can call more than one function, in a sequence, whenever it is triggered.

Ingredients:

- Python3
- A text editor (preferably an IDE with Python support)
- The Pyo module
- the wxPython module, if you want to use Pyo's GUI
- A terminal window in case your editor does not launch Python scripts

Process:

An example of the use of `Score()` is shown in Script 3.27. The first argument of this class must be a PyoObject that outputs integers. These will be used to determine the function to be called. The second argument is a part of the name of a function to be called. This part of the name will be concatenated with the integer number passed to the first argument to assemble the name of a function. For example, if we want to call the functions `func_0()`, `func_1()`, and `func_2()` in a row, the second argument to `Score()` must be "`func_`".

Script 3.27 The Score() class.

```
 1 from pyo import *
 2
 3 s = Server().boot()
 4
 5 adsr=Adsr(attack=.04,decay=.08,sustain=.6,release=.06,dur=.2)
 6 fm = FM(carrier=200, ratio=.248, index=adsr*8, mul=adsr*0.3)
 7 mix = Mix(fm.mix(1), voices=2).out()
 8
 9 def func_0():
10     fm.setCarrier(200)
11     fm.setIndex(adsr*4)
12     adsr.play()
13
14 def func_1():
15     fm.setCarrier(250)
16     fm.setIndex(adsr*6)
17     adsr.play()
18
19 def func_2():
20     fm.setCarrier(300)
21     fm.setIndex(adsr*8)
22     adsr.play()
23
24 metro = Metro(time=.2).play()
25 counter = Counter(metro, min=0, max=3)
26 score = Score(counter, fname="func_")
27
28 s.gui(locals())
```

The first six lines of the code in Script 3.27 are almost identical to the ones in Script 3.26. In this code though we want to control some attributes of the `FM()` object, and if we call its `mix(2)` method, when we try to call one of its methods that controls its attributes we will get an `AttributeError` saying that the 'Mix' object has no such attribute. To remedy this, we create a `Mix()` object explicitly, like we have been doing in previous examples.

In lines 9, 14, and 19 we define three functions with the same name and an incrementing integer concatenated to their names. These functions set a frequency to the `carrier` attribute, a value to the `index` attribute, and they activate the envelope. Note that we initialise our envelope with a duration of 0.2 seconds, so we don't need to call its `stop()` method. We set the attributes of the `FM()` object by calling the corresponding method of the attribute we want to change, instead of calling `fm.carrier=200`.

In line 24 we create a metronome with the `Metro()` class. In this class we define the time intervals between each trigger and we call its `play()` method to activate it. `Metro()` will output trigger signals at these time intervals. A trigger signal in Pyo is a sample of 1 surrounded by 0s. More on samples further on in this chapter. In line 25 we initialise our counter. This time the counter is not a simple integer, like it was in Script 3.26, but a PyoObject. This object takes a PyoObject that sends trigger signals as its first argument, and a minimum and a maximum value (which is excluded from the count).

Finally in line 26 we initialise our `Score()` object where we pass our counter to its first argument, and the string "`func_`" to its `fname` attribute. `fname` stands for "function name". Once we start our script, our three functions will be called sequentially. Make sure the duration of the envelope is at least as long as the accumulation of the values of its `attack`, `decay`, and `release` attributes, otherwise you will get audible clicks.

Documentation Pages:

`Metro()`: https://belangeo.github.io/pyo/api/classes/triggers.html#metro
`Counter()`: https://belangeo.github.io/pyo/api/classes/triggers.html#counter
`Score()`: https://belangeo.github.io/pyo/api/classes/pattern.html#score

3.4.2.3 Metro() and TrigEnv()

`Pattern()` and `Score()` are good for triggering functions that don't need sample accuracy, like the examples in Scripts 3.26 and 3.27, but when we need timing to be very tight, we have to look elsewhere, and that is in the sample accurate timing classes of Pyo. In Script 3.28 we create a simple drum machine based on audio samples. If we were to create this with any of the event sequencing classes, we would get a lot of jitter when we raised the tempo. That is because standard Python functions that are called by Pyo's event sequencing classes are affected by the rather limited timing accuracy of Python. Being an interpreted language, not needing compiled binary files to run, Python is considered to be a rather slow language, and when it comes to strict temporal constraints, using standard Python functions may not be the best choice.

Ingredients:

- Python3
- A text editor (preferably an IDE with Python support)
- The Pyo module
- the wxPython module, if you want to use Pyo's GUI
- A terminal window in case your editor does not launch Python scripts

Process:

We will create a 16-step drum machine with a kick drum, a snare drum, and a hi-hat. We will set the triggering state of each step through a standard Python list, where a 1 triggers and a 0 does not trigger.

Script 3.28 A simple drum machine based on audio samples.

```
1 from pyo import *
2
```

```
 3 s = Server().boot()
 4
 5 sample_paths = ["./samples/kick-drum.wav",
 6                 "./samples/snare.wav",
 7                 "./samples/hi-hat.wav"]
 8 # create a three-stream SndTable
 9 snd_tabs = SndTable(sample_paths, chnl=0)
10 # get a list of the durations
11 durs = snd_tabs.getDur(all=True)
12
13 # create triggering lists
14 kick_list  = [1, 0, 0, 0, 1, 0, 0, 0, 1, 0, 0, 0, 1, 0, 0, 0]
15 snare_list = [0, 0, 1, 0, 0, 0, 1, 0, 0, 0, 1, 0, 0, 0, 1, 0]
16 hat_list   = [0, 1, 1, 0, 1, 1, 0, 1, 1, 0, 1, 1, 0, 1, 1, 0]
17 # put all lists in one list
18 all_lists = [kick_list, snare_list, hat_list]
19
20 # get ms from BPM based on meter and pass it to Metro's arg
21 metro = Metro(beatToDur(1/4, 120))
22 mask = Iter(metro, all_lists)
23 # create a dummy arithmetic object
24 beat = metro * mask
25
26 # create a three-stream TrigEnv()
27 player = TrigEnv(beat, snd_tabs, dur=durs, mul=.5).out()
28
29 def new_list(l, index=0):
30     all_lists[index] = l
31     mask.setChoice(all_lists)
32
33 metro.play()
34
35 s.gui(locals())
```

In line 5 we create a list of strings with the path to the audio files we will be using. These are written in Unix syntax used by Linux and macOS. The dot means the current directory – the directory from where the script is run – and the forward slash separates directories and files from one another. For this script to work, you must create a directory (a folder) named "samples", and in that directory you must place your sound files. If their names are different, you must set them properly in this line. If you don't have any drums samples, you can find a lot at freesound.org.

In line 9 we create an SndTable(). This class stores an audio file to a table that can later be used by a sample accurate timing class. By passing a list to its first argument, we create a multi-stream object of this class. We set the chnl attribute to 0, to read the first channel of each audio file. If we omit this, the diffusion of the triggering and the output channels in the resulting sound will be messy. In line 11 we store a list with the durations of the sound files, by calling the getDur() method of the SndTable() class with the kwarg all set to True. The default value of this kwarg is False, and if we leave it like that, we will get the duration of the first audio file only, as a float, not a list.

In lines 14 to 16 we create the lists that set the triggering state of each of the sixteen steps of our sequencer. We set the kick drum to hit on the down beats, the snare drum to hit on the up beats, and a pattern that repeats every three beats for the hi-hat. In line 16 we place the three lists to another list. We do this to be able to change a list while the script is running.

In line 21 we create our metronome. Since we are building a drum machine it makes more sense to express time in Beaps Per Minute (BPM) instead of seconds, which is `Metro()`'s time unit. Pyo includes a function that converts BPM to seconds, given the division of the beat as an argument too. We want our sequencer to run at 120 BPMs, but we want to trigger 16 notes in a 4/4 bar, so we need 1/4 of the beat duration.

Line 22 creates an `Iter()` object. This object receives trigger signals in its first argument, and iterates over a list that is set in its second argument. In our case we pass a list of lists, so we create three audio streams of this class. Line 24 creates what in Pyo is called a dummy arithmetic object. Pyo enables the user to make arithmetic operations on PyoObjects and the result will be a dummy object that runs at the audio rate and its output is the result of the arithmetic operation. In this line, we multiply the trigger signals of the metronome by the lists stored in the `mask` object of the `Iter()` class. Remember that trigger signals are a value of 1 when the triggering occurs, surrounded by 0s. If we multiply this by the values in our lists, then whenever a list has a 1, the triggering signal will go through, and whenever it has a 0 the triggering signal will not go through. We can think of these lists as a gate for the triggers. Due to this masking, we name the `Iter()` object `mask`.

In line 27 we create a `TrigEnv()` object. This object takes a trigger signal, and whenever it receives a trigger, it reads values from a table at the duration passed to its `dur` kwarg. Since the second and third arguments are lists, this object will also have three audio streams, one for each audio sample. In line 33 we call the `play()` method of the `Metro()` class to start our metronome.

What is left to explain is the `new_list()` function, in line 29. This function can be called while the script is running. It takes one list with sixteen items, and an optional kwarg that sets the index of the audio stream, 0 for the kick drum, 1 for the snare, and 2 for the hi-hat. The list we will pass will replace the respective list from lines 14 to 16. Once the list in `all_lists` has been replaced, we call the `setChoice()` method of the `Iter()` class, with this list of lists as the argument. For example, the following line sets all 16 steps of the hi-hat to active:

```
new_list([1 for i in range(16)], index=2)
```

We can also change the tempo by calling the `setTime()` method of the `Metro()` class. This can either happen in two stages where we will first store the BPM to milliseconds by calling the `beatToDur()` function, or we can pass this function straight to the `setTime()` method. If you do the latter, make sure you have the correct number of open and closed round brackets. For every bracket that opens, one must close it.

You will notice that this script alternates the audio streams to a stereo setup. The kick drum and hi-hat are heard through the left channel, and the snare drum through the right. We will see how to handle output channels diffusion further down in this chapter. Also notice that Script 3.28 includes some comments, explaining what some lines do. This is a standard in computer programming, and a habit you should start to get the hang of.

Documentation Pages:

`TrigEnv()`: https://belangeo.github.io/pyo/api/classes/triggers.html#trigenv
`Iter()`: https://belangeo.github.io/pyo/api/classes/triggers.html#iter
Pyo's sample accurate timing classes: https://belangeo.github.io/pyo/api/classes/triggers.html

3.4.3   MIDI

Since Pyo is a music focused Python module, it could not omit MIDI support. MIDI is a communication protocol used between hardware and software. It was initialised in the 1980s and it is still the most popular communication protocol for controlling electronic instruments. Its name stands for Musical Instrument Digital Interface. Even though MIDI 2.0 has already been launched, MIDI 1.0 is still the most prominent protocol used. Initially MIDI messages were transferred through a 5-pin DIN cable, but since many years, MIDI messages are usually transmitted through USB cables. In this section we will examine how we can receive Note and Control MIDI messages.

Ingredients:

- Python3
- A text editor (preferably an IDE with Python support)
- The Pyo module
- the wxPython module, if you want to use Pyo's GUI
- A MIDI keyboard (preferably including potentiometers)
- A terminal window in case your editor does not launch Python scripts

Process:

To run the code of this section you will need a MIDI keyboard and a MIDI controller with knobs. Most of MIDI keyboards have knobs, in which case the keyboard controller will suffice. The first thing we need to do is check what number our MIDI device has. Conveniently enough, Pyo includes a function that prints all available MIDI devices in your system. The code in Script 3.29 does this.

Script 3.29 Printing available MIDI devices on the console.
```
1 from pyo import *
2
3 s = Server().boot()
4 s.start()
5
6 pm_list_devices()
```

When you run this script, you will get a list with a number, the mode of the device (input or output), and the name of the device. We don't need to run this script in a loop, but only once, so we don't use the `gui(locals())` method of Pyo's `Server()`, but we start the server by calling its `start()` method in line 4. Once you know the number of your device, you can use that as an argument in the `setMidiInputDevice()` method of the `Server()` class. For control messages we need to know the controller number and channel. Script 3.30 duplicates the example from the documentation of the `CtlScan()` class.

Script 3.30 Scanning the MIDI control devices on your system.
```
1 from pyo import *
2
3 s = Server()
4
5 s.setMidiInputDevice(5)
```

```
 6 s.boot()
 7
 8 def ctl_scan(ctlnum):
 9     print(ctlnum)
10
11 a = CtlScan(ctl_scan)
12
13 s.gui(locals())
```

Note that to set a MIDI device, we must first create the server without booting it. Once we set our MIDI device, we can then boot the server. If you run this code and you have a MIDI device that sends control messages connected to your computer, when you turn its potentiometers, you will see a list printed to the console that looks like this:

```
ctl number : 21, ctl value : 60, midi channel : 1
```

In our case, we need the first and last values. Script 3.31 uses all this information to control a CrossFM() class. This is similar to FM(), but both oscillators modulate the frequency of one another, which means that the modulator modulates the carrier, and the carrier modulates the modulator. The only additional attribute this class has, is the second index. The two index attributes are named ind1 and ind2.

Script 3.31 Controlling a CrossFM() synth with MIDI.

```
 1 from pyo import *
 2
 3 s = Server()
 4 s.setMidiInputDevice(3)
 5
 6 s.boot()
 7
 8 notes = Notein(poly=10, scale=1, mul=.5)
 9 adsr=MidiAdsr(notes['velocity'],attack=.005,decay=.1,
10               sustain=.4, release=1)
11 ratio = Midictl(21, channel=1, mul=.5)
12 index1 = Midictl(22, channel=1, mul=5)
13 index2 = Midictl(23, channel=1, mul=5)
14 xfm = CrossFM(carrier=notes['pitch'], ratio=ratio,
15               ind1=index1, ind2=index2, mul=adsr).out()
16
17 s.gui(locals())
```

In line 8 we initialise a Notein() object that will receive the NoteOn and NoteOff MIDI messages. These messages contain two values, one for the note, and one for its velocity. The velocity expresses how hard a key on the keyboard is pressed. MIDI range is 7-bit, so all MIDI values go from 0 to 127 inclusive. A maximum velocity will have the value 127, but the Notein() class takes care of scaling it to a range between 0 and 1 for us.

Documentation Pages:

CtlScan(): https://belangeo.github.io/pyo/api/classes/midi.html#ctlscan
Notein(): https://belangeo.github.io/pyo/api/classes/midi.html#notein

MidiAdsr(): https://belangeo.github.io/pyo/api/classes/midi.html#midiadsr
Midictl(): https://belangeo.github.io/pyo/api/classes/midi.html#midictl
CrossFM(): https://belangeo.github.io/pyo/api/classes/generators.html#crossfm
Pyo's MIDI classes: https://belangeo.github.io/pyo/api/classes/midi.html

3.4.4   Routing and Panning

We have already briefly seen how we can send audio to a specific speaker, or how to send a stereo signal. In this section we will clarify certain concepts concerning sound routing, and we will see how we can apply panning to our scripts.

3.4.4.1   Sound Routing

In Pyo we can choose the output channel (essentially, which loudspeaker) for an audio signal by passing a numeric argument to its out() method. Playing in stereo though can be achieved in a few different ways.

Ingredients:

- Python3
- A text editor (preferably an IDE with Python support)
- The Pyo module
- the wxPython module, if you want to use Pyo's GUI
- A terminal window in case your editor does not launch Python scripts

Process:

Script 3.32 is a very simple example that outputs a sine wave oscillator at 440Hz to the left speaker, and another sine wave oscillator at 660Hz, to the right speaker. If we want to send a single audio signal to both channels of a stereo setup, we can call the mix(2) method. We must bear in mind though that this method will change the audio object from whatever class it is assigned to (Sine() in this example) to the Mix() class.

Script 3.32 Simple setup with two sine waves with 3/4 frequency ratio.
```
1 from pyo import *
2
3 s = Server().boot()
4
5 sineL = Sine(freq=440, mul=.2).out()
6 sineR = Sine(freq=660, mul=.2).out(1)
7
8 s.gui(locals())
```

Script 3.33 creates a single sine wave oscillator and sends it to both speakers, but immediately after its initialisation, it calls its setFrequency() method, to change the frequency. This throws an AttributeError stating that the "'Mix' object has not attribute 'setFreq'". We have already seen how to compensate for this. We can send our signal to a Mix() object instead. Then we can call any method of the oscillator without getting errors.

Script 3.33 The mix(2) method causing an AttributeError.

```
1 from pyo import *
2
3 s = Server().boot()
4
5 sine = Sine(freq=440, mul=.2).mix(2).out()
6 sine.setFreq(660)
7
8 s.gui(locals())
```

Another way to send a single audio source to two or more channels is by passing a list to at least one of the kwargs of a class. Script 3.34 does what Script 3.33 attempts to do, only this script succeeds.

Script 3.34 Outputting in stereo without getting errors.

```
1 from pyo import *
2
3 s = Server().boot()
4
5 sine = Sine(freq=[440, 440], mul=.2).out()
6 sine.setFreq([660, 660])
7
8 s.gui(locals())
```

3.4.4.2 *Panning*

Panning sound is the action of "moving" sound from one side to the other. When panning sound in a stereo setup, the two speakers will have the same volume only when we want the sound to be centred. In all other occasions, the volume between the two speakers will differ. The simplest formula to set the volume to the two speakers is the following linear formula:

```
Ramp = 1 - Lamp
```

Ramp is the amplitude of the right speaker, and *Lamp* is the amplitude of the left. In this scenario, when we want to centre the sound, both speakers will have an amplitude of 0.5. This way of controlling amplitude though makes us perceive the sound farther away when it is centred, than when it is positioned at one or the other end (Roads, 1996, p. 459). You can imagine an elliptical movement between the two speakers, that stretches farther away than a circular movement. Figure 3.18 illustrates this, where the circle at the bottom is the listener, the other two, smaller circles are the speakers, and the grey straight lines are the distance between the listener and the speakers. The grey curve is the curve of equal power panning, where the distance at any point in the curve is equal to that in any of the two ends. The black curve is the curve of linear panning. We can see that it appears to be farther away than what it is expected to be.

The amplitude value the speakers should get to properly place a sound in the middle is 0.707, which is the square root of 0.5. To avoid doing all the calculations when we want to move sound in space, Pyo includes two classes for panning, Pan() and SPan(). Pan() is a cosinus panner that includes a spread factor. This factor determines how much sound will leak to the surrounding channels. SPan() is a simple equal power panner. Script 3.35 uses the Pan() class

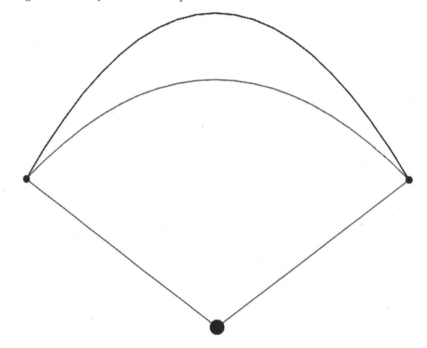

Figure 3.18 Linear panning vs constant power panning.

to pan white noise in a stereo setup. The `outs` attribute sets the number of channels `Pan()` will pan the sound to. The `pan` and `spread` attributes take values between 0 and 1. The `spread` attribute though is ineffective in a stereo setup. It is functional only on a multi-channel setup.

Note that using `Pan()` or `SPan()` is necessary only if you are moving sound in your speaker setup. If your sounds do not move, then these classes are not necessary and you can send your audio signals straight to your speakers with the amount of amplitude you like.

Script 3.35 Panning white noise in a stereo setup.

```
 1 from pyo import *
 2
 3 s = Server().boot()
 4
 5 noise = Noise(mul=.2)
 6 pan = Pan(noise, outs=2, pan=.5, spread=0).out()
 7
 8 pan.ctrl()
 9
10 s.gui(locals())
```

Documentation Pages:

`Pan()`: https://belangeo.github.io/pyo/api/classes/pan.html#pan
`SPan()`: https://belangeo.github.io/pyo/api/classes/pan.html#span

3.5 Sampling Rate, Bit Depth, Buffer Size, Latency, and Audio Range

Now that we have built a foundation on electronic sound, it is time to discuss some technical aspects of digital audio. Specifically, in this section, we will discuss sampling rate, buffer size, and audio range.

3.5.1 Sampling Rate and Bit Depth

Sampling rate is the speed at which the sound is sampled. In digital audio, the resulting sound for each output channel is a series of discrete amplitude values, which are lowpass filtered in the internal or external audio interface, to create a smooth audio signal (Roads, 1996, p. 24). The most common sampling rate is 44,100Hz also expressed as 44.1kHz. This is the sampling rate used in CD quality. This means that for one second of sound, we need 44,100 discrete amplitude values. Figure 3.19 illustrates an example of a sine wave with discrete amplitude values that are used in digital audio.

The sampled values in Figure 3.19 are usually within the audio range, meaning that the highest value will be 1 and the lowest -1. These values are converted to voltage inside the circuit of the sound card, and after this is passed through the filtering circuitry that "connects the dots", it is sent to the output. The alternating voltage ends up in the speakers and moves their woofers back and forth, with the higher values pushing the woofer forward, and the lower values pulling it backward.

The higher the sampling rate, the clearer the sound. We can get a better idea of how sampling rate affects audio if we think that telephone audio is sampled at 8kHz. This very low rate is enough for us to understand who we are talking to and what our interlocutor is saying. When it comes to music and electronic sound though, the demands are much higher than those of phone calls. Another typical sampling rate is 48kHz. Higher sampling rates are also possible. 96KHz,

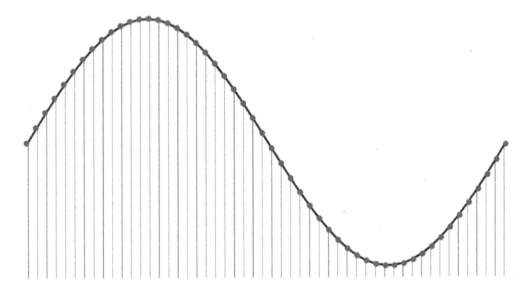

Figure 3.19 A sine wave sampled at regular intervals.

or even 192kHz are sampling rates that can be found in hardware and software. It is arguable though whether applying such high rates has any useful effect. A 48kHz rate should be sufficient for any electronic music application.

The bit depth determines the signal-to-noise ratio, also known as SNR. It also determines the dynamic range of our signal. Standard CD quality bit depth is 16 bits. This means that each sample of the sound we want to record or produce, can be expressed in binary numbers with 16 bits. This bit depth covers a range of 65,536 discrete values. This means that the -1 to 1 audio range can be expressed with a 65,536 resolution (Roads, 1996, p. 930). Each bit we add to our resolution doubles the previous range. 17 bits cover a range of 131,072. This doubling provides a gain of approximately 6dB in the dynamic range. A 16-bit system can have a theoretical range of 96dB (Roads, 1996, p. 930), whereas 20 bits provide a 120dB range, which is the resolution of the human ear. Most modern software has a resolution of 64 bits, and most modern sound cards have 24 bits.

3.5.1.1   Aliasing

An effect of sampled digital audio is aliasing, also known as foldover. This occurs when we try to represent a sound with a frequency higher than the highest possible frequency, based on our sampling rate. With a 48kHz rate, we can represent sounds up to 24kHz frequencies, half our sampling rate. This is because we need at least one amplitude value for the highest part of one period of our sound's waveform, and one for the lowest. This frequency threshold is called the Nyquist frequency (Yao, 1983), named after the physicist Harry Nyquist.

When we try to exceed this frequency, the sampled points we get start to become less than two per cycle, resulting in shifting the pitch of the sound. Figure 3.20 illustrates a waveform reproduced at a frequency higher than the Nyquist frequency. The black outline is the waveform attempted to be produced, the grey dots are the sampled points, the grey square wave is the resulting digital waveform, and the grey sine wave is the sound output by the audio interface after filtering.

The resulting frequency we get when aliasing is yielded by the following equation: Nyquist frequency – (frequency – Nyquist frequency). If we attempt to output a sine wave at 25kHz, with a 48kHz sampling rate, then the actual frequency will be 24 – (25–24) = 23kHz. Essentially, the resulting frequency is folding over at the Nyquist frequency and starts dropping as much as we exceed it. So, if we exceed it by 100Hz, we'll hear the Nyquist frequency minus 100. This is why this effect is also called foldover. The default sampling rate in Pyo is 44.1kHz, but it can be set to any sampling rate supported by your soundcard via the `sr` attribute, or the `setSamplingRate()` method of the `Server()` class.

3.5.2   Buffer Size and Latency

Having to compute thousands of values per second is a very demanding task, even for fast computers. With sampling rates of 44.1kHz or higher, every audio process we add to our programs adds to this task. To lighten the burden of the vast number of calculations, digital audio is processed in blocks of samples. When doing audio processing, the computer will process all the samples in one block, as fast as it can, and it will then send them to the sound card. For real-time audio, the computer must be capable of computing faster than the time it takes for one sample block to sound. A rather small sample block size, or buffer size, is 64 samples. These sizes are expressed in values of 2 to some power. At a 48kHz sampling rate, a sound of 64 samples takes approximately 1.33333 milliseconds (ms). This means that the computer must process these

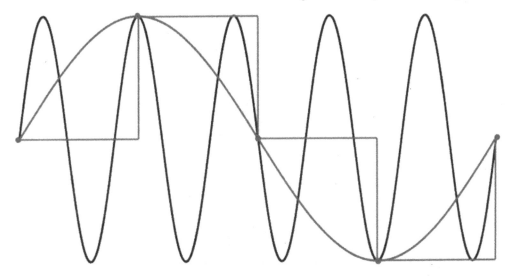

Figure 3.20 An aliased sine wave.

samples faster than this tiny duration, otherwise we will get audible clicks, also known as drop-outs. Once these samples have been processed, the computer can do something else, if there is time. When the time it takes for these samples to sound is over, the DSP clock will notify our program to compute the next block. This clock runs at a frequency specified by the equation: 1000 / (sampling rate/sample block size), which yields the duration of the sample block in ms.

Modern computers can easily process this number of samples faster than the time it takes to hear them. The more we drop the size of the sample block though, the more the time difference between the time it takes the computer to process them and the time these samples take to sound is reduced. So, the more we drop our sample block size, the less time the computer has for other tasks. Depending on the priority of a task, it is likely that our program will be pushed back in the task list, and we can get dropouts if we drop the size of the sample block very low. Systems with low-latency or real-time kernels provide a very high priority to audio tasks, and it is very unlikely that you can get dropouts. Modern systems though are fast enough to not need such a kernel for most applications. Throughout this book we will not be doing anything that requires such fast processing so that a low-latency or real-time kernel will be necessary.

Big block sizes produce what is called latency. If we set our block size to 2,048, the frequency of the DSP clock will be approximately 42.7ms. The threshold of sound perception is 20ms (Gabor, 1946), so such a big block size will make it very easy for the computer to process the sound, but it will also produce a big latency that will very likely be audible, especially if we process audio input. Setting the block size is a trade-off between minimising the CPU load and a responsive system. The default block size in Pyo is 256 and it is set via the `buffersize` attribute, or the `setBufferSize()` method of the `Server()` class. If you are using the Jack audio server, then these attributes are set in Jack, and not Pyo, as Pyo will follow whatever settings Jack has.

3.5.3 *Audio Range*

We have already discussed the audio range, but we need to discuss what happens when we exceed it. In digital audio, values greater than 1 and smaller than -1 are not possible to be output

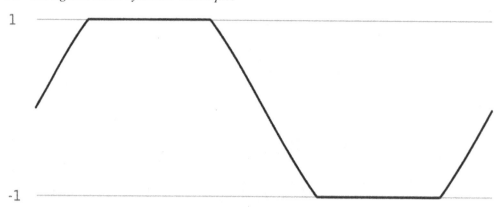

Figure 3.21 A clipped sine wave.

through the hardware. If we exceed these values in the waveforms we send to the DAC,[2] these waveforms will be clipped. Figure 3.21 illustrates a clipped sine wave. Clipping adds harmonics to sound, as it adds sharp corners to the waveform. Even if this is a desired effect, it should be done programmatically, and not through the limitations of the hardware. The effect of Figure 3.21 can be achieved with Pyo's `Clip()` class.

3.6 FFT

The Fast Fourier Transform (FFT) is an efficient implementation of the Discrete Fourier Transform (DFT) algorithm (Roads, 1996, p. 551). These algorithms take their names from the French mathematician and physicist Jean-Baptiste Joseph Fourier. They are used to analyse the spectrum of a sound and decompose it to its sinusoidal components. The DFT and FFT analyses are perceived to represent a spectral "snapshot" of a sound (Boulanger and Lazzarini, 2011, p. 521). They transform a sound from the time domain to the frequency domain. One of the most popular analysis and resynthesis techniques that incorporate FFT is the Phase Vocoder (Roads, 1996, p. 549) which we will see in the next chapter.

Ingredients:

- Python3
- A text editor (preferably an IDE with Python support)
- The Pyo module
- the wxPython module, to use Pyo's GUI
- A terminal window in case your editor does not launch Python scripts

Process:

Pyo includes a number of classes that perform an FFT analysis, resynthesis, or other functions on the analysis data. Since it includes an even greater number of Phase Vocoder classes, in this section we will briefly see how we can analyse, process, and resynthesise a signal by applying the FFT. The code in Script 3.36 applies an FFT analysis on white noise, and then filters its spectrum, before it resynthesises it.

Script 3.36 Applying a spectral filter to white noise with FFT.

```
 1 from pyo import *
 2 import random
 3
 4 s = Server().boot()
 5
 6 def create_list():
 7     # create a list of ten random integers
 8     # from 0 to 512 without repetitions
 9     rand_idx = random.sample([i for i in range(512)], 10)
10     # create a list of tuples with the random integers
11     # and a random float
12     rand_list = [(rand_idx[i], random.random())
13                     for i in range(len(rand_idx))]
14     # sort the list in ascending order
15     rand_list.sort()
16     list_w_zeros = []
17
18     for i in range(len(rand_list)):
19         # if the index of the first item
20         # of the list is greater than 0
21         if i == 0 and rand_list[i][0] > 0:
22             # go half way between index and beginning
23             index = int(rand_list[i][0] / 2)
24             # add a 0 value to half way index
25             list_w_zeros.append((index, 0))
26         # if successive indexes are separated
27         # with index difference greater than 1
28         elif rand_list[i][0] - rand_list[i-1][0] > 1:
29             # go half way between indexes
30             index=int((rand_list[i][0]-rand_list[i-1][0])/2)
31             # add offset to place index at correct spot
32             index += rand_list[i-1][0]
33             # add a zero value to that index
34             list_w_zeros.append((index, 0))
35         # add the tuple of the original list
36         list_w_zeros.append(rand_list[i])
37     # add a list tuple between last item and end of table
38     if rand_list[-1:][0][0] < 511:
39         index = int((512 - rand_list[-1:][0][0]) / 2)
40         index += rand_list[-1:][0][0]
41         list_w_zeros.append((index, 0))
42     return list_w_zeros
43
44 list_w_zeros = create_list()
45
46 # create a list from a set() to remove duplicates
47 no_rep = [*set(list_w_zeros)]
48 # sort list as set() shuffles items
49 no_rep.sort()
50
51 noise = Noise(.25).mix(2)
```

```
52 fft = FFT(noise, size=1024, overlaps=4, wintype=2)
53 tab = ExpTable(no_rep, size=512)
54 amp = TableIndex(tab, fft["bin"])
55 real = fft["real"] * amp
56 imag = fft["imag"] * amp
57 ifft = IFFT(real, imag, size=1024, overlaps=4,
58            wintype=2).mix(2).out()
59
60 tab.graph()
61
62 s.gui(locals())
```

Before we discuss the create_list() function, we will discuss the FFT part of the script, from line 51 onward. In line 51 we create a stereo stream of white noise, and in line 52 we create an FFT() object. The first argument is the signal to analyse. The second argument is the number of samples the object will use for the analysis. The FFT takes a sound from the time domain and transfers it to the frequency domain. To do that, we need to obtain a certain number of samples. This number will provide us with bins that represent frequency bands. With a 1024 size, which is the default, we will be splitting the range of our sampling rate to 1024 bands. Each band will get a value that will denote the amount of energy found in that frequency band. This way, we can decompose a signal and get the energy it contains in each analysis bin. This number must be 2 raised to the power of some value, here it is raised to the 10th power.

The overlaps attribute sets the number of overlapped analysis blocks. The greater this number, the better the analysis and resynthesis, but it affects the CPU load. This value must be a positive integer, and it defaults to 4. The wintype attribute sets the type of the envelope used to filter each frame. The FFT() class provides nine choices, from no windowing, to eight different window types. The default is the "Hanning" window, which we have already seen in the Granular Synthesis section, and it is indexed with the number 2.

In line 53 we create an ExpTable() object. This class creates a table where we can set values based on indexes, and it will apply an exponential interpolation between these values. Its first argument is a list of tuples with the index and its value. We create a random list programmatically with the create_list() function, and we set the size to be half our FFT size. This is because the second half of the FFT analysis is over the Nyquist frequency and we don't care about that.

In line 54 we create a TableIndex() object. Here we index the ExpTable() with the "bin" part of the FFT() object. The FFT() class outputs three audio streams, the "real" part, the "imaginary" part, and the "bin" part. The FFT analysis decomposes a signal to a series of complex values with a real and an imaginary part, expressed with cosine and sine components respectively. Each pair of cosine and sine values represent a frequency point in our frequency bins, within the analysis. Both parts are necessary for the resynthesis of a signal. The "bin" part of the FFT output is simply the bin index of the analysis. In line 54 we use this incrementing index to read values from the ExpTable().

Lines 55 and 56 is where the spectral processing is done. What we do is multiply both the real and imaginary part of the FFT analysis by the values in the ExpTable(), bin by bin. Figure 3.22 illustrates this table, initialised with random values. Imagine this figure spanning the audible range from 0Hz to its left side up to the Nyquist frequency to its right. If we superimpose a grid with 512 separate regions onto this image, then we can see the amplitude of the analysed signal in each bin illustrated by the exponential lines.

Finally, in line 57 we resynthesise our spectrally filtered signal with the IFFT() class. IFFT stands for Inverse-FFT. It takes both the real and imaginary parts of the FFT analysis, and the rest of the arguments of the FFT() class, which should have the same values. In line 60 we call the graph() method of the ExpTable() class, to display the table shown in Figure 3.22. This widget is interactive and you can move the dots by clicking and dragging with the mouse, or you can create new dots by clicking on a spot where there is no dot. Experiment with the dots and listen to the sound change.

The last part we need to discuss is the create_list() function. Comments have been intentionally placed in almost all lines in this function so the reader can understand what is happening only by reading the code. What this function does is create a list of tuples with random indexes and random values. The indexes should not repeat, and to achieve that, we use the random.sample() method in line 9, to get a list with 10 random integers from 0 to 512, that don't repeat. Then we create a list with ten tuples with these non-repetitive random values and a random float between 0 and 1, using list comprehension. Note that we encapsulate rand_idx[i] and random.random() in parenthesis. We do this because this is the correct Python syntax to create a tuple. Since the non-repetitive random integers will not be sorted, we need to sort then in ascending order, and we do this in line 15.

The loop in line 18 adds tuples with indexes half way between the existing random indexes of the rand_list with values of 0. If this is omitted, the table in Figure 3.22 would not include the dots that touch the bottom of the image, and all the other dots would get connected, resulting in a much noisier sound, since its spectrum would filter much less frequency bins. In line 38 we check if the last random index is not 511, which is the maximum index, and if it is true, we add another 0 value half way between the last random index and the end of the table. Finally, in line 42 we return the list with the added 0s, and in line 44 we call this function and store the list it returns to list_w_zeros. Note that there is a local list inside the function with the same

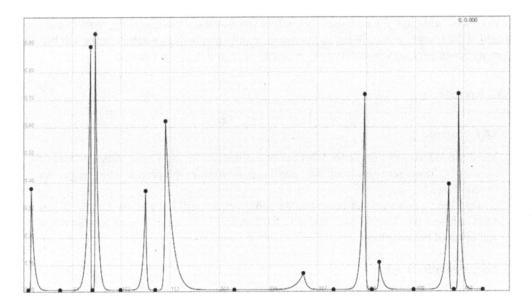

Figure 3.22 Random points in an ExpTable() object.

name, initialised in line 16. Since the scope of the two variables is different, and the global list is initialised after the function definition, there is no name clash.

It is possible to get duplicate tuples with zeros, if two random indexes are separated by only two integers apart. To remedy this, in line 47 we use the `set()` data type we saw in Chapter 2, to eliminate this possibility. We pass the list with all the 0s included as its argument, and we encapsulate it in square brackets to retrieve a Python list out of it. We need to place the asterisk symbol to get the elements of the `set()` unpacked, otherwise our list would contain a single element, the `set()` itself. Finally we sort this list too, as the `set()` data type shuffles its contents.

Documentation Pages:

`FFT()`: https://belangeo.github.io/pyo/api/classes/fourier.html#fft
`ExpTable()`: https://belangeo.github.io/pyo/api/classes/tables.html#exptable
`IFFT()`: https://belangeo.github.io/pyo/api/classes/fourier.html#ifft

3.7   Conclusion

In this chapter we took a deep dive into many audio synthesis techniques. We started from the most fundamental element of electronic sound, the oscillator, and moved on to utilise this, together with other tools, to create sound. From RM, AM, FM, Additive and Granular Synthesis, we moved on to effects, filters, sequencing, panning, up to one of the most advanced topics in electronic music, the FFT. By now you should have a good understanding of how electronic sound is produced and controlled, and you are now ready to move on and start creating musical applications. The chapters that follow will build on Chapters 2 and 3, to create meaningful musical programs, with the complexity increasing more or less linearly.

At this point it is a good time to join the Pyo community in its discussion panels. The main portal is Pyo's Google group.[3] At the time of writing there is also a Discord channel named "PYO users", although for now there are only a few members there, and the main discussion is held at the Google group. Being an open source software, both Pyo and its users will benefit from its community, so you are strongly encouraged to join it, if you use Pyo.

3.8   Exercises

3.8.1   *Exercise 1*

Make the square wave oscillator from Script 3.2 with the `Compare()` class, instead of `Round()`. Make sure you bring the resulting waveform to the proper audio range, between -1 and 1.

Tip: `Compare()` cannot be sent to the DAC, so you will have to use `Sig()` if you want to hear it too. Use the `mul` and `add` attributes of `Compare()` to correct the range and offset of the waveform.

Documentation Page:

`Compare()`: https://belangeo.github.io/pyo/api/classes/utils.html#compare

3.8.2 Exercise 2

Change Script 3.7 so that the modulator signal is added to the carrier frequency outside of the car object. Use a `Scope()` to compare your version against that of Script 3.7.

Tip: Use a `Sig()` class, but don't do the addition the same way it is done in Script 3.7. Instead, use `Sig()`'s attributes to achieve the same result.

3.8.3 Exercise 3

Using `HarmTable()` and `Osc()`, create an impulse waveform. Use list comprehension to create your list of partial amplitudes.

Tip: All the partials in this waveform have an amplitude of 1. Use the `autoNormalize()` method to correct the range of the waveform.

3.8.4 Exercise 4

Replace the hard-coded value of the `pitch` attribute of the `Granulator()` and `Granule()` objects in Script 3.18 with an LFO. Make sure to scale the output of the LFO to a small range and to give it the right offset so its output is around the hard-coded values in Script 3.18.

Tip: Even though Pyo includes an `LFO()` class, any Pyo oscillator can be an LFO, if its frequency is in the infra-audio range. Take care to create two streams in your oscillator's output, by passing a list to one of its kwargs, since the `pitch` attribute takes a list in Script 3.18. Use the `mul` and `add` attributes to scale and offset the oscillator, respectively.

Documentation Page:

Pyo's signal generators: https://belangeo.github.io/pyo/api/classes/generators.html

3.8.5 Exercise 5

Change the envelope in Script 3.25 to one with curved lines.

Tip: You will need the `Expseg()` class to achieve this. This class takes a list of tuples with two values, the time and the value at that point in time. For example, `[(0,0),(.5,1),(1,0)]` will create an envelope that starts at 0, goes to 1 in half a second, and drops down to 0 again after 1 second since it started. To imitate an ADSR envelope you will need to start at 0, go to 1 in a short time, drop to a lower value, around 6.5, in a short time again, but a bit longer than the attack, stay there for the longest duration of the envelope, and drop to 0 in a short time again.

An example of time stamps could be 0, 0.1, 0.3, 1, 1.25. An envelope with these time stamps would last 1.25 seconds, it would rise to 1 in 0.1 seconds, drop to the sustain value in 0.2 seconds, stay at that value for 0.7 seconds, and its release stage would last 0.25 seconds. Note that to maintain the sustain value you must define two time stamps with the sustain value, its beginning and its end. The exponent factor should be below 1 to get the right curves. A value of 0.5 should work well. You should set the `inverse` attribute to `False`. Use the `graph()` method of the class to visualise the envelope.

Documentation Page:

`Expseg()`: https://belangeo.github.io/pyo/api/classes/controls.html#expseg

3.8.6 Exercise 6

Write a script that does exactly what Script 3.27 does, but use `Pattern()` instead of `Score()`.

Tip: Create two lists, one with the carrier frequencies, and one with the coefficients multiplied by the `adsr` object, passed to the index of the `FM()` class. Use a global variable in the calling function to iterate over these lists. Make sure you don't exceed the length of the lists, otherwise you will get an `IndexError`. Remember that lists are zero-based indexed.

3.8.7 Exercise 7

Change Script 3.28 to output all audio samples to both channels in a stereo setup.

Tip: Route the output sound to a `Mix()` object the same way we have done it in previous examples. Make sure the, implicitly created, input `mix()` has the correct number of channels, and the `Mix()` object has the correct number of voices.

3.8.8 Exercise 8

Use Script 3.31 without a hardware MIDI keyboard.

Tip: You can use Pyo's GUI keyboard. You should not set a MIDI device to the server, and not create the `Midictl()` objects. You must hard-code the `ratio`, `ind1`, and `ind2` attributes, since you will not have a way to control them. Call the `keyboard()` method of the `Notein()` object, but make sure you have first created this object. A GUI keyboard will pop up when you run the script. You can use this keyboard with your mouse, but also with your computer's keyboard. The following keys are active:

q = middle C, 2 = C#, w = D, 3 = D#, e = E, r = F, 5 = F#, t = G, 6 = G#, y = A, 7 = Bb, u = B, i = C, 9 = C#, o = D, 0 = D#, p = E

If you picture this, it is like a piano keyboard, with the q-w-e-r-t-y-u-i-o-p keys being the white keys, and 2-3-5-6-7-9-0 being the black keys. The following keys are one octave lower, similarly to the ones above (the hyphens are used to separate the characters, they are not active in the keyboard): z – s – x – d – c – v – g – b – h – n – j – m – , – l – . – ; – /

Notes

1 To get the analysis to look like Figure 3.11, you have to drag the horizontal bar at the bottom of the spectrum window from its right corner toward the left.
2 Digital to Analog Converter
3 https://groups.google.com/g/pyo-discuss

Bibliography

Bjørn, K. and Meyer, C. (2018) *Patch and Tweak*. Denmark: Bjooks.

Boulanger, R. and Lazzarini, V. (2011) *The Audio Programming Book*. Cambridge, Massachusetts: MIT Press.

Chowning, J. (1973) 'Synthesis of complex audio spectra by means of frequency modulation', *AES: Journal of the Audio Engineering Society*, 21, pp. 526–534.

Cipriani, A. and Giri, M. (2009) *Electronic Music and Sound Design*. Rome: ConTempoNet s.a.s.

Farnell, A. (2010) *Designing Sound*. Cambridge, Massachusetts: MIT Press.

Gabor, D. (1946) 'Theory of communication. Part 1: The analysis of information', *Journal of the Institution of Electrical Engineers - Part III: Radio and Communication Engineering*, 93, pp. 429–441.

Huovilainen, A. and Välimäki, V. (2005) 'New approaches to digital subtractive synthesis', in *Proceedings of the International Computer Music Conference, ICMC*. Barcelona, Spain, pp. 300–402.

Lazzarini, V. *et al.* (2016) *Csound*. Springer. Available at: https://doi.org/10.1007/978-3-319-45370-5.

Rabiner, L. *et al.* (1972) 'Terminology in digital signal processing', *IEEE Transactions on Audio and Electroacoustics*, 20(5), pp. 322–337. Available at: https://doi.org/10.1109/TAU.1972.1162405.

Roads, C. (1996) *The Computer Music Tutorial*. Cambridge, Massachusetts: MIT Press.

Smith, J.O. (2010) *Physical Audio Signal Processing*. https://ccrma.stanford.edu/~jos/pasp/Freeverb.html.

Wilson, S., Cottle, David and Collins, N. (2011) *The SuperCollider Book*. Cambridge, Massachusetts: MIT Press.

Xenakis, I. (1992) *Formalized Music, Thoughts and Mathematics in Composition*. Bloomington: Indiana University Press.

Yao, Y.-C. (1983) 'Nyquist Frequency', *NASA STI/Recon Technical Report N* [Preprint]. Available at: https://doi.org/10.1002/0471667196.ess1840.pub2.

4   Phase Vocoder Techniques

In this chapter we will experiment with the Phase Vocoder (PVOC). Even though initially developed for non-musical reasons (Roads, 1996, p. 549), the PVOC has been used extensively in Computer Music applications, with the first experiments dating back to the 1970s (Moorer, 1976).

What You Will Learn

After you have read this chapter you will:

- Know what the PVOC is and its various applications
- Know how to easily apply cross synthesis between two sound sources
- Know how to easily morph between two audio spectra
- Know how to easily filter the spectrum of a sound
- Know how to easily apply spectral reverberation
- Become acquainted with the various Pyo PVOC classes

4.1   What is the Phase Vocoder?

The PVOC is a tool for spectral analysis, processing, and resynthesis of sounds with FFT at its heart. It takes its name from the Vocoder, which is a contraction of the term "voice encoder" or "voice coder" (Dolson, 1986, p. 14). In the electronic music world, it was initially used to alter the duration or pitch of a sound, independently of one another, while the sound remained unaltered when no modifications were made (Moorer, 1976). In its early days, the PVOC was used with recorded sounds, transforming them in musical ways (Dolson, 1986, p. 23). The PVOC is now used with arbitrary input signals and operates on their spectra in various ways, not only aiming to transform pitch or duration, but other parameters too.

Pyo includes an ecosystem of classes that perform various operations of audio spectra, including frequency transposition, cross synthesis between audio inputs, spectral filtering, spectral morphing, amplitude and frequency modulation, and others. This ecosystem is based on the `PVAnal()` and `PVSynth()` classes, where the former makes the analysis, and the latter the resynthesis of sound. In the simplest case, a sound can be analysed and resynthesised to get the original sound unaltered. The interesting part though, is when we process the analysis stream output by `PVAnal()` before we resynthesise it with `PVSynth()`.

DOI: 10.4324/9781003386964-4

In this chapter we will look at cross synthesis, spectral morphing, spectral additive synthesis, spectral filtering, and spectral reverberation. We have already seen the spectral filtering in the previous chapter, but in this chapter we will see how Pyo enables us to perform this task with minimal coding.

4.2   Cross Synthesis

Pyo's `PVCross()` class performs cross synthesis on two audio inputs. This class takes the PVOC analysis of two `PVAnal()` objects and applies a mix of their amplitudes scaled by a factor, to the amplitudes of the first input.

Ingredients:

- Python3
- A text editor (preferably an IDE with Python support)
- The Pyo module
- the wxPython module, if you want to use Pyo's GUI
- A terminal window in case your editor does not launch Python scripts

Process:

`PVAnal()` outputs two streams, similarly to an FFT analysis, but these outputs are the magnitudes – referred to as amplitudes in `PVCross()` – and the frequencies of the analysed sound. What `PVCross()` does, is to take the amplitudes of both analysed signals, mix them, and replace the original amplitudes of its first input with this mix. An example that highlights the effect of cross synthesis is the mixture of a noisy conversation with a harmonic sound. This effect will likely sound familiar to you when you run the code in Script 4.1.

Script 4.1 Cross synthesis between a noisy conversation and a harmonic sound.

```
1 from pyo import *
2
3 s = Server().boot()
4
5 sf1 = SfPlayer("samples/harm_sound.wav", loop=True, mul=.5)
6 sf2 = SfPlayer("samples/noisy_conv.wav", loop=True, mul=.5)
7 pva1 = PVAnal(sf1)
8 pva2 = PVAnal(sf2)
9 pvc = PVCross(pva1, pva2, fade=1)
10 pvs = PVSynth(pvc).out()
11
12 pvc.ctrl()
13
14 s.gui(locals())
```

`PVAnal()` takes a PyoObject as its argument. The user can also change other parameters, like the FFT size, which defaults to 1024, the number of overlaps, which defaults to 4, the window type, which defaults to Hanning, and a callback function. The last kwarg sets a function that is called at the end of every analysis and its arguments must be the magnitudes and frequencies of the analysis. This can be useful if you want to test the analysis data. As you can see, all the kwargs but the last, are the same as the FFT() kwargs. We will keep the default values for all these, so we don't need to add them to our code.

In lines 5 and 6 we create two SfPlayer() objects where we load two samples, a noisy conversation and a harmonic sound. The freesound.org website is a good place to get some audio samples if you don't have any. Then in lines 7 and 8 we make a PVOC analysis on each of these audio files. In line 9 we create our PVCross() object where we pass the two PVOC analyses as its first two arguments, and we also set the fade kwarg to 1, its default value.

The fade attribute of PVCross() is the factor that scales the amplitudes of the two sounds. A 1 means that the first input of PVCross() will get the amplitudes of the second input, and a 0 means that it will get its own amplitudes, hence the signal will be resynthesised unaltered. To get the familiar effect of the human voice sounding like a synthesizer, the first input to PVCross() must be the harmonic sound. This means that we need to hear the amplitudes of the noisy sound with the frequencies of the harmonic sound.

To resynthesise the signal, we must send the output of PVCross() to a PVSynth() object. This object takes a PVOC stream of magnitudes and frequencies and resynthesises it back to the time domain. This happens in line 10, where we also output the sound to the speakers. If your audio files are stereo, then you will get a stereo output. If they are mono, then you will get sound only from the first output channel. Use the "fade" slider of the "PVCross" widget to hear how this attribute affects the sound.

It has already been mentioned that PVOC was initially used on sound samples and that now it is used with any kind of input. You can replace the two SfPlayer() objects with input from a microphone that you get with Input(), or with an oscillator. Mind to place the analysis of a harmonic sound to the first argument of PVCross(), whatever sound sources you use. Of course, inverting the sound sources in PVCross()'s arguments is also interesting, and you are encouraged to experiment with this.

You should note that if you want to change any parameter of the PVAnal() objects, other than the window type, you should do it to both objects simultaneously, as PVCross() needs its input PVOC analysis streams to have the same FFT size and the same number of overlaps. You can either do this by creating a function that changes any of these two parameters to both PVAnal() objects, or use channel expansion in the arguments of PVAnal(). Script 4.2 defines two functions, each changing one of these two attributes of both PVAnal() objects.

Script 4.2 Functions to change the parameters of all PVAnal() objects simultaneously.
```
1 def size(x):
2     pva1.setSize(x)
3     pva2.setSize(x)
4
5 def olaps(x):
6     pva1.setOverlaps(x)
7     pva2.setOverlaps(x)
```

Documentation Pages:

PVAnal(): https://belangeo.github.io/pyo/api/classes/pvoc.html#pvanal
PVCross(): https://belangeo.github.io/pyo/api/classes/pvoc.html#pvcross
PVSynth(): https://belangeo.github.io/pyo/api/classes/pvoc.html#pvsynth

4.3   Spectral Morphing

The `PVMorph()` class applies spectral morphing between two PVOC input streams. It is very similar to `PVCross()` in the way a `PVMorph()` object is initialised, but it is different in the processing it does on the PVOC streams.

Ingredients:

- Python3
- A text editor (preferably an IDE with Python support)
- The Pyo module
- the wxPython module, if you want to use Pyo's GUI
- A terminal window in case your editor does not launch Python scripts

Process:
The `PVMorph()` class takes two `PVAnal()` objects and applies spectral morphing between both amplitudes and frequencies, in contrast to `PVCross()` that morphs between the amplitudes only. It takes a `fade` kwarg that sets the interpolation between the amplitude and frequency vectors of the `PVAnal()` objects, where this interpolation is linear for the amplitudes and exponential for the frequencies. Script 4.3 is copied from the documentation page of `PVMorph()`. It morphs between an oscillator and a loop of an audio file with a human voice.

Script 4.3 Spectral-morphing between an oscillator and a human voice.
```
 1 from pyo import *
 2
 3 s = Server().boot()
 4
 5 sl = SineLoop(freq=[100,101], feedback=0.12, mul=.5)
 6 sf = SfPlayer(SNDS_PATH+"/transparent.aif", loop=True, mul=.5)
 7 pva1 = PVAnal(sl)
 8 pva2 = PVAnal(sf)
 9 pvm = PVMorph(pva1, pva2, fade=.5)
10 pvs = PVSynth(pvm).out()
11
12 pvm.ctrl()
13
14 s.gui(locals())
```

`SineLoop()` is a sine wave oscillator that modulates its own phase. It feeds its output back to its phase, with the `feedback` kwarg setting how much of this feedback signal will modulate its phase. The `SfPlayer()` object plays the voice sample that comes with Pyo. In line 9 we create our `PVMorph()` object and pass the two `PVAnal()` objects to its first two inputs. Since we call its `ctrl()` method, we will get a widget to control the `fade` attribute while the script is running. Remember that the only object from the PVOC ecosystem that can be sent to the output is `PVSynth()`, with the only exception of `PVAddSynth()` that we will see next.

As with Script 4.1, in Script 4.3 we must be careful if we want to change the size or the number of overlaps of the PVOC analysis. This should happen to both `PVAnal()` objects, like it is done in Script 4.2. If you want to experiment with these attributes, copy the code from Script 4.2 and paste it into Script 4.3.

Documentation Pages:

SineLoop(): https://belangeo.github.io/pyo/api/classes/generators.html#sineloop
PVMorph(): https://belangeo.github.io/pyo/api/classes/pvoc.html#pvmorph

4.4 PVOC Additive Synthesis

Pyo's PVOC ecosystem provides a class that recreates a PVOC stream with Additive Synthesis. This is the PVAddSynth() class, the only class after PVSynth() from PVOC that can be sent to the output.

Ingredients:

- Python3
- A text editor (preferably an IDE with Python support)
- The Pyo module
- the wxPython module, if you want to use Pyo's GUI
- A terminal window in case your editor does not launch Python scripts

Process:

PVAddSynth() provides an easy way to transpose the pitch of a sound (although the same effect can be achieved with PVTranspose()) and to set the number of oscillators used for the Additive Synthesis to recreate a sound. This class takes a PVOC stream from any of the PVOC classes and uses the magnitudes and frequencies of the stream to control the amplitude and frequency envelopes of an oscillator bank. Script 4.4 is copied from PVAddSynth()'s documentation page.

Script 4.4 Spectral resynthesis with Additive Synthesis with PVAddSynth().
```
 1 from pyo import *
 2
 3 s = Server().boot()
 4
 5 sf = SfPlayer(SNDS_PATH+"/transparent.aif",loop=True, mul=0.7)
 6 pva = PVAnal(sf)
 7 pvs = PVAddSynth(pva, pitch=1.25, num=100, first=0, inc=2)
 8 mix = Mix(pvs.mix(1), 2).out()
 9
10 s.gui(locals())
```

The interesting part in PVAddSynth() is in its kwargs. The pitch kwarg controls the amount of transposition, with 1 resynthesising the sound unaltered, values higher than 1 shifting the pitch upward, and lower than 1 shifting it downward. The num attribute controls the number of oscillators that will be used to resynthesise the sound. The less the oscillators the more lowpass filtered the resynthesised audio will sound like. first sets the first bin in the analysis to start the resynthesis from. Remember that in an FFT analysis, which is at the heart of PVOC, a bin represents the energy in a frequency band. The higher this value, the more bins from the lower part of the analysis will be left outside of the resynthesis, resulting in the audio sounding like it is highpass filtered. The `inc` kwarg sets every how many bins the resynthesis will happen, starting from bin `first`. The greater this number, the sparser the bins in the resynthesis will be and the more altered the audio will sound.

Documentation Page:

`PVAddSynth()`: https://belangeo.github.io/pyo/api/classes/pvoc.html#pvaddsynth

4.5 Spectral Filtering with PVOC

We have already seen how we can apply filtering to the spectrum of a sound, at the end of Chapter 3. In this section we will see how we can achieve the same thing with the PVOC ecosystem.

Ingredients:

- Python3
- A text editor (preferably an IDE with Python support)
- The Pyo module
- the wxPython module, if you want to use Pyo's GUI
- A terminal window in case your editor does not launch Python scripts

Process:

Script 4.5 results in the exact same process as Script 3.36 from Chapter 3, where we did an FFT analysis on white noise and used an `ExpTable()` to control the energy in each bin of the analysis. The steps in Script 4.5 are fewer though, as the `PVFilter()` class and the PVOC ecosystem make the process of spectral analysis, processing, and resynthesis much easier than using raw FFT classes.

Script 4.5 Applying spectral filtering with PVFilter().

```
1  from pyo import *
2  import random
3
4  s = Server().boot()
5
6  def create_list():
7      rand_idx = random.sample([i for i in range(512)], 10)
8      rand_list = [(rand_idx[i], random.random()) for i in
9                      range(len(rand_idx))]
10     rand_list.sort()
11     list_w_zeros = []
12
13     for i in range(len(rand_list)):
14         if i == 0 and rand_list[i][0] > 0:
15             index = int(rand_list[i][0] / 2)
16             list_w_zeros.append((index, 0))
17         elif rand_list[i][0] - rand_list[i-1][0] > 1:
18             index=int((rand_list[i][0]-rand_list[i-1][0])/2)
19             index += rand_list[i-1][0]
```

```
20                    list_w_zeros.append((index, 0))
21              list_w_zeros.append(rand_list[i])
22        if rand_list[-1:][0][0] < 511:
23            index = int((512 - rand_list[-1:][0][0]) / 2)
24            index += rand_list[-1:][0].[0]
25            list_w_zeros.append((index, 0))
26        return list_w_zeros
27
28  list_w_zeros = create_list()
29
30  no_rep = [*set(list_w_zeros)]
31  no_rep.sort()
32
33  noise = Noise(.25).mix(2)
34  pva = PVAnal(noise)
35  tab = ExpTable(no_rep, size=512)
36  pvf = PVFilter(pva, tab)
37  pvs = PVSynth(pvf).out()
38
39  tab.graph()
40
41  s.gui(locals())
```

The comments in the `create_list()` function have been removed, since this function is identical to the one from Script 3.36 in Chapter 3, and it was discussed there. We should pay attention to the lines 34 onward where we apply spectral filtering with PVOC. In line 34 we create the PVOC analysis object and in line 35 we create the `ExpTable()` object with the non-repetitive list passed to its first argument. The size is again 512, since the default size of the PVOC analysis is 1024.

To apply spectral filtering with PVOC we don't need to retrieve the real and imaginary part from the analysis, nor do we need to iterate through the `ExpTable()` with the `TableIndex()` class, like we did in Script 3.36. We simply pass the PVOC analysis and the `ExpTable()` object to the `PVFilter()` object and we are done. We then pass the PVOC streams of our `PVFilter()` object to `PVSynth()` and it does the spectral filtered resynthesis.

Since PVOC has an FFT at its heart, many effects of Pyo's PVOC ecosystem can be achieved with the FFT classes as well. The PVOC ecosystem though, makes these processes a lot easier, with cleaner and less code.

Documentation Page:

`PVFilter()`: https://belangeo.github.io/pyo/api/classes/pvoc.html#pvfilter

4.6   Spectral Reverberation

The last PVOC technique we will see is the spectral reverberation. This is achieved with the `PVVerb()` class from the PVOC ecosystem.

Ingredients:

- Python3
- A text editor (preferably an IDE with Python support)

- The Pyo module
- the wxPython module, if you want to use Pyo's GUI
- A terminal window in case your editor does not launch Python scripts

Process:

The `PVVerb()` class applies reverberation in the spectral domain. Script 4.6 uses this class to apply reverb to the voice sample that comes with Pyo. Notice that in this script we use a larger size in the PVOC analysis than what we have used up to now. The size plays a dramatic role in the sound quality of `PVVerb()`. This attribute affects other PVOC classes as well, but in `PVVerb()` the sound difference between different sizes is very prominent. Run this script and change the size of `pva` while the script is playing. The size should be a power of 2, so try values like 1024, 2048, 4096, and 8192. The entire process will start from the beginning, so the reverberation will clear and will start again. You should hear very different results with different values. Note though that the greater the size, the greater the latency, as it takes more time to collect 4,096 samples than the time it takes to collect 1,024 samples. That should not be a problem with an audio sample like the one used in Script 4.6, but if you want to apply spectral reverberation to live audio input, then the latency will start to become audible.

Script 4.6 Spectral reverberation with PVVerb().

```
 1 from pyo import *
 2
 3 s = Server().boot()
 4
 5 sf = SfPlayer(SNDS_PATH+"/transparent.aif", loop=True, mul=.5)
 6 pva = PVAnal(sf, size=2048)
 7 pvv = PVVerb(pva, revtime=0.95, damp=0.95)
 8 pvs = PVSynth(pvv).mix(2).out()
 9
10 pvv.ctrl()
11
12 s.gui(locals())
```

Documentation Page:

`PVVerb()`: https://belangeo.github.io/pyo/api/classes/pvoc.html#pvverb

4.7   Conclusion

We have been introduced to the Phase Vocoder and to the PVOC ecosystem of Pyo. This set of objects provides an easy and intuitive way to apply PVOC effects to any sound, not just for pitch shifting, but for many more effects. We saw that even though we can achieve the same results by using the FFT classes of Pyo, it is easier to use the PVOC classes, as we need to write less code resulting in cleaner programs that are easier to understand. You are encouraged to check the rest of the PVOC classes and their examples, and see what this versatile ecosystem is capable of.

Documentation Page:

Pyo's PVOC classes: https://belangeo.github.io/pyo/api/classes/pvoc.html

4.8   Exercises

4.8.1   Exercise 1

Modify Script 4.3 so that one input is a `SineLoop()` oscillator and the other input is your voice from your systems audio input with `Input(0)`. Use a MIDI keyboard to control the `SineLoop()` oscillator so that your voice sounds like a synthesizer. Determine the order of the inputs to get a result similar to that of Script 4.3.

   Tip: Use `Notein()` and `MidiAdsr()` to control the oscillator, like we did in Script 3.31 in Chapter 3. If you don't have a MIDI keyboard, use the `keyboard()` method of the `Notein()` class.

4.8.2   Exercise 2

Use the FFT classes to recreate Script 4.3.

   Tip: Make an FFT analysis of each sound source and convert its output from Cartesian to Polar coordinates. Then mix the Polar coordinate outputs to convert back to Cartesian before you resynthesise the signal with the Inverse Fast Fourier Transform.

Documentation Page:

Pyo's FFT classes: https://belangeo.github.io/pyo/api/classes/fourier.html

4.8.3 Exercise 3

Modify Exercise 2 by replacing the classes that do a Cartesian to Polar conversion and vice versa with trigonometric functions.

   Tip: To get the magnitudes of the noisy sound, you will need the hypotenuse where the known values are the real and imaginary outputs of the FFT analysis. To get the angles of the harmonic sound, you will need the arc tangent of the real and imaginary outputs of the FFT analysis.

   To convert back to polar coordinates, for the real part, you will need the cosine of the angles which then you must multiply by the magnitudes. For the imaginary part, you will need the sine of the angles, and then again multiply it by the magnitudes. Then you can resynthesise with the Inverse Fast Fourier Transform.

Documentation Page:

Pyo's arithmetic classes: https://belangeo.github.io/pyo/api/classes/arithmetic.html

Bibliography

Dolson, M. (1986) 'The Phase Vocoder: A Tutorial', *Computer Music Journal*, 10, p. 14.
Moorer, J.A. (1976) 'The Use of the Phase Vocoder in Computer Music Applications', *Journal of the Audio Engineering Society*, 26(9), pp. 42–45.
Roads, C. (1996) *The Computer Music Tutorial*. Cambridge, Massachusetts: MIT Press.

5 Controlling Pyo with Physical Computing

In this chapter we will connect Pyo to the physical world by connecting various sensors to the computer and getting their data into Python and Pyo. For many scenarios, using MIDI controllers can be sufficient for controlling various parameters, but being able to use any sensor we want in our Python scripts, provides another level of flexibility and freedom.

What You Will Learn

After you have read this chapter you will:

- Know how to connect an Arduino to Python
- Know how to get data from sensors into Python
- Know how to send data from Python to the Arduino
- Know how to use sensor data musically with Pyo
- Be able to complete a project with a custom circuit and your own code

---

5.1 What is Physical Computing

Physical Computing is a term that encompasses various Human-Computer Interaction (HCI) practices through the use of sensors that convert analogue signals to digital, transferring information from the physical world to the computer, and vice versa. Through Physical Computing, we are able to get information such as proximity, acceleration, vibration, and many others, into our programs and do anything we want with this. Once digitised, information can be transformed to anything else digitised. With this approach, we can use digitised signals of any sensor in any way we like in a Python script. In this chapter we will focus on the use of the popular Arduino, and how we can use it musically combined with Pyo.

5.1.1 What is the Arduino

The Arduino is a prototyping platform consisting of a hardware micro-controller board, its programming language and IDE, and its community (Drymonitis, 2015). The Arduino board comes in different versions that cover a wide range of project requirements. In this chapter we will focus on the most fundamental version, the Arduino Uno. This board, shown in Figure 5.1, is a prototype-friendly board that enables fast wiring with minimal to no soldering, so the users can have their projects up and running in a short time.

DOI: 10.4324/9781003386964-5

Figure 5.1 An Arduino Uno.

The language used to program the Arduino is C++ with a wide range of functions comprising its Application Programming Interface (API). These functions provide an intuitive interface for getting input from sensors and providing output to LEDs, motors, solenoids, and other elements. The Arduino IDE is the official IDE used to program the board. It provides functionalities like auto-completion, colour-highlighting, auto-indentation, and other features found in IDEs. The most important features of this IDE though, is the interface for uploading code to an Arduino board, its debugger, and the serial monitor and plotter. These features make programming the Arduino easy, and provide all the necessary tools to develop Physical Computing projects.

The last part of the Arduino is its community. The Arduino has a large international community of engineers, hobbyists, tinkerers, and enthusiasts that share knowledge and projects through the various communication channels, like the official forum and others. The official website of the Arduino is https://www.arduino.cc/. It includes all the necessary information to get you started, as well as links to its various community channels.

5.2   How Can We Connect the Arduino to Python

The standard way to connect the Arduino to the computer is to program the former in its language and connect the program running on the computer to communicate with the USB port the Arduino physically connects to. Then we can retrieve the data using the communication protocol of the Arduino. This approach though requires the use of two programming languages, Arduino and any language we program the computer in. It is possible to control the Arduino through other programming languages by using the Firmata firmware. This firmware includes code for the Arduino that the user uploads to the board once. Then it is possible to send information using native commands of another programming language and either send or request data from the Arduino.

By using the Firmata firmware, we can use pure Python to communicate with the Arduino and we don't have to worry about the Arduino code. The downside is that requesting data from the Arduino is a two-step process, as we first have to send a command to the Arduino and then wait for the response with the requested data. This two-step process adds overhead to the intercommunication, and sometimes it adds latency that can become perceivable. For the purposes of this chapter, this overhead is not problematic, so we will use both approaches to the communication between the computer and the Arduino. Before we move on and connect the Arduino to Python, you will need to install the Arduino IDE to your computer. Head over to www.arduino.cc/en/software and install the latest version of the IDE.

5.2.1 Testing Our Arduino Board

The first thing we need to do is to ensure we can upload code to the Arduino board. Launch the Arduino IDE and go to File → Examples → 01.Basic → Blink. In Arduino jargon, the code of a program is called a sketch. Clicking on the "Blink" menu item, the "Blink" sketch will open in a new window. This is the most fundamental micro-controller program, similar to the "Hello World!" program we wrote in Chapter 2, but for electronics.

Ingredients:

• An Arduino Uno board
• The Arduino IDE

Process:

In Sketch 5.1 we can see the "Blink" code. What this sketch does is turn an LED on for one second, then turn it off for another second, and repeat this process indefinitely. The philosophy in electronics is that, if we can light up an LED, we can do anything, so this test ensures we will be able to use the hardware and software of our micro-controller. Providing a detailed explanation on the Arduino code is beyond the scope of this chapter and book, so we will touch the Arduino sketches briefly, as we will be focusing on the Python code.

Sketch 5.1 The "Blink" Arduino sketch.

```
1 void setup() {
2     pinMode(LED_BUILTIN, OUTPUT);
3 }
4
5 void loop() {
6     digitalWrite(LED_BUILTIN, HIGH);
7     delay(1000);
8     digitalWrite(LED_BUILTIN, LOW);
9     delay(1000);
10 }
```

The Arduino programs run the two functions we see in Sketch 5.1. The setup() function is run once when the board is powered on, and the loop() function runs repeatedly immediately after `setup()` is done. The `void` keyword at the function definition lines declares that the following function will not return anything. In C++ we must declare all the data types of all variables, as well as the data type that is to be returned by a function, explicitly. If one of these functions were to return an integer, instead of `void`, we would have to write `int`. In the `setup()` function of the "Blink" sketch, all we need to do is declare the mode of the pin we want to use, by calling the `pinMode()` function. The Arduino has both analog and digital pins to which we can connect analog or digital sensors and receive or send data. Figure 5.2 highlights these pins, where the top pins are the digital, and the bottom are the analog.

The "Blink" sketch uses the digital pin 13 because it includes a built-in LED, hence the `LED_BUILTIN` macro[1] used in line 2. This macro is defined as the number 13 in Arduino's source code. The second macro used in the `pinMode()` function is the `OUTPUT`, which tells the function to set this pin as an output pin. Arduino can use all its digital pins either as input or output.

Once the `setup()` function is done, the `loop()` function will run repeatedly. In this function we raise the voltage of the digital pin 13 high, sending 5 Volts (5V), then wait for 1000 milliseconds (ms), drop the voltage of the pin to 0V, and wait for another 1000ms. The native Arduino functions for these actions are self-explanatory. `digitalWrite()` writes a voltage value to a pin, taking two arguments, the pin number and the voltage value. The `HIGH` macro is equivalent to the integer 1. This sets the voltage of a digital pin high. The `delay()` function is similar to Python's `time.sleep()`. It stalls the program for a certain amount of time, only its argument is the time expressed in ms, rather than seconds.

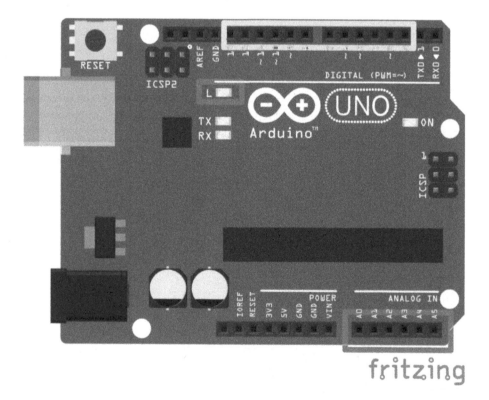

Figure 5.2 An Arduino Uno with its digital and analog pins, and the on-board LED highlighted.

In the Arduino IDE go to Tools → Board → Arduino AVR Boards, and select Arduino Uno (or whichever other board you are using). Then go to Tools → Port, and select the serial port of the Arduino. The drop-down list should state which port the Arduino is connected to, by appending an "Arduino Uno" in parenthesis. On Linux, this is usually */dev/ttyACM1*, on Windows it is *COM1*, and on Mac it is */dev/cu.usbmodemXXX* (X is some integer). Once you have setup your board and port, click on the "Upload" button, the button with the arrow. If all is well, you should see the upload progress (it should be short), and once the code is uploaded to your board, you should see the on-board LED light up and turn off. This LED is highlighted in Figure 5.2.

5.2.2   *Connecting the Arduino to Python*

Now that we have tested our board, we can move on and connect the Arduino to Python and run the "Blink" sketch from there.

Ingredients:

- Python3
- The pySerial module
- The pyFirmata module
- A text editor (preferably an IDE with Python support)
- An Arduino Uno
- The Arduino IDE
- A terminal window in case your editor does not launch Python scripts

Process:

We will first start with the pySerial module that communicates with the serial lines of the computer with raw bytes. We will use this to replicate the "Blink" sketch. Then we will do the same with the pyFirmata module.

5.2.2.1   *Using pySerial*

With pySerial we must program the Arduino explicitly. Sketch 5.2 shows the Arduino code. For this code to run with Python, you will need the pySerial Python module. Type `pip install pyserial` in a terminal window and the module should install. If `pip` refers to Python2, write `pip3` to install pySerial for Python3.

Sketch 5.2 Arduino code to control the on-board LED from Python.

```
 1 void setup() {
 2     pinMode(LED_BUILTIN, OUTPUT);
 3     Serial.begin(115200);
 4 }
 5
 6 void loop() {
 7     if (Serial.available()) {
 8         int val = Serial.read() - '0';
 9         digitalWrite(LED_BUILTIN, val);
10     }
11 }
```

Our `setup()` function now includes a call to the `Serial.begin()` function. This function starts the serial communication so we can send and receive data between the computer and the Arduino. Its argument is what is called the baud rate. This is essentially the number of bits per second that will go through the USB connection. We set this to a rather high value, 115,200. In the `loop()` function, we check if there is data waiting in the serial buffer by calling `Serial.available()`. This function returns the number of bytes that are waiting in the serial buffer. If the number of data is greater than 0, the conditional test will succeed and its code will be executed. In this code, we read one byte from the serial line with `Serial.read()`.

We will be sending bytes in the form of ASCII[2], so a 1 will be sent as the byte 49, and a 0 will be sent as 48. To get the actual value we need, we have to subtract 48 from this byte, which is also expressed with the 0 character. To define this as a single character, we enclose it in single quotes. Think of it as a one-character string whose integer value is 48, the ASCII value of 0. Once we get the value we want, we write it to our digital pin, so the on-board LED can light up or turn off. Upload this sketch to your Arduino board. By default, when uploading a sketch, it is also saved, so the IDE will prompt you to save it to some location in your computer, which is likely to be ~/Arduino (the tilde character means the home directory). When saving an Arduino sketch, it is automatically saved in a directory with the same name as the sketch. To control the on-board LED from Python, write the code in Script 5.1 and run it.

We must first import the pySerial module and then we must open the Arduino port by creating a `Serial()` object and passing the Arduino port name and the baud rate as creation arguments. Then we can send a 1 and a 0 to control the on-board LED. Note that we must send data in the `bytes` data type of Python. We do this by writing a b, and then the value we want, in single or double quotes.

Script 5.1 Controlling the on-board LED of the Arduino with Python and pySerial.

```
1 import serial
2 from time import sleep
3
4 ser = serial.Serial("/dev/ttyACM1", 115200)
5
6 while True:
7     ser.write(b'1')
8     sleep(1)
9     ser.write(b'0')
10    sleep(1)
```

5.2.2.2   *Using pyFirmata*

Now let's try the same with the Firmata firmware. Arduino inlcudes the code for Firmata by default. In the Arduino IDE, go to File → Examples → Firmata → Standard Firmata. A rather long sketch will open in a new window. Upload this to your Arduino board and you are done with the Arduino side of things. In Python you must install the pyFirmata module by typing `pip install pyfirmata`. Script 5.2 replicates the functionality of Script 5.1, with the use of the Firmata firware.

Script 5.2 Controlling Arduino's on-board LED from Python with Firmata.

```
1 from pyfirmata import Arduino
2 from time import sleep
3
```

```
 4 board = Arduino("/dev/ttyACM1")
 5
 6 while True:
 7     board.digital[13].write(1)
 8     sleep(1)
 9     board.digital[13].write(0)
10     sleep(1)
```

Note that we don't need to specify the baud rate when we initialise an `Arduino` object. This is because Firmata has a fixed baud rate of 57,600 and that is also hard-coded in pyFirmata. It also seems that we don't need to set the mode of the pin, as the code in Script 5.2 works without any such call. The rest of the code should be self-explanatory.

When running any of these three scripts, in the beginning you will see the on-board LED blink fast a few times. This happens while a connection between Python and the Arduino is being established. Also, when the connection has been established and the `while` loop starts running, together with the on-board LED, you will see another LED next to it blink shortly when a command is sent from Python to the Arduino. This is the serial receive LED, indicating that data has been received in the serial buffer.

5.3 Using Proximity Sensors to Control Amplitude and Frequency

Now that we have established a connection and we have sent data from Python to the Arduino, it is time to receive data from the Arduino to Python. We will be using an infra-red proximity sensor and use its data to control the amplitude of an oscillator. Then we will add another proximity sensor to control the frequency, simulating a Theremin.

5.3.1 Using One Proximity Sensor to Control the Amplitude of an Oscillator

We will start with one proximity sensor that will control the amplitude of a sine wave oscillator. We will do this with both approaches to the communication between Python and the Arduino.

Ingredients:

- Python3
- The Pyo module
- The wxPython module, if you want to use Pyo's GUI
- The pySerial module
- The pyFirmata module
- A text editor (preferably an IDE with Python support)
- An Arduino Uno
- The Arduino IDE
- One infra-red proximity sensor
- A terminal window in case your editor does not launch Python scripts

Process:

An infra-red sensor detects proximity from an obstacle by sending pulses of an infra-red beam and counting the time it takes to receive it back. When an obstacle is in front of the sensor, the beam will reflect and bounce back to the sensor. By calculating how long it took for the beam to bounce back, the sensor adjusts its output voltage to express this distance. The Arduino reads this voltage and converts it from analog to digital in a 10-bit resolution. This means that analog values from the Arduino have a range between 0 and 1023, 1024 discrete values in total. With an infra-red sensor though, usually this range is smaller.

What we need to do first is check the maximum value this sensor sends, so we can convert its full range to a range between 0 and 1, so we can properly control the amplitude of an oscillator. For that, we will use the Arduino IDE only, without any Python involved. Before we write the Arduino code, we must create our circuit. This is shown in Figure 5.3. The proximity sensor has three colour-coded wires coming out of it. The red wire connects to the power source, which is the 5V pin on the Arduino, the black connects to ground, which is the GND pin on the Arduino, and the yellow (or some other colour) connects to the analog pin 0. These wires are exposed at their end, so we can connect them straight to the pin headers of the Arduino. Just twist the wires at the end and slide them in the headers as it is shown in Figure 5.3.

Sketch 5.3 shows the test code. What happens in this code should be straight forward. We initialise the serial communication with a 115,200 baud rate and then we read the first analog pin, indexed with 0, by calling the `analogRead()` function, with the pin number as its argument. We then print the value to the serial line by calling the `println()` function of the Serial class. This class has two functions that print data onto the serial line, `print()` and `println()`. The former prints anything that is passed to its argument. The latter prints its argument and appends the carriage return and newline characters. The return character is the character sent

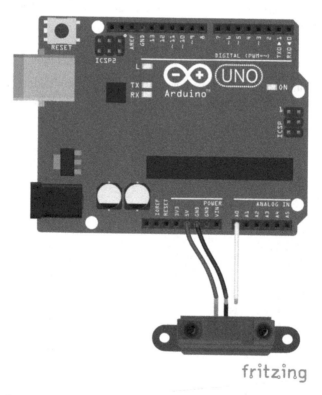

Figure 5.3 Infra-red proximity sensor circuit.

when we press the Return key on the keyboard. The newline character is the character sent when we want to create a new line in a program. Their ASCII bytes are 13 and 10, respectively, and their characters are expressed with `'\r'` and `'\n'`, respectively. If we print the values with `print()` instead of `println()`, every value will be concatenated to the previous one when printed. With `println()`, after we print a value, we force the serial monitor of the Arduino to go one line below, so each value will be printed in its own line.

Sketch 5.3 Proximity sensor test sketch.

```
1 void setup() {
2     Serial.begin(115200);
3 }
4
5 void loop() {
6     int infraval = analogRead(0);
7     Serial.println(infraval);
8     delay(500);
9 }
```

Upload this code to your Arduino board and click on the Serial Monitor icon, on the top right part of the Arduino IDE (the icon with the magnifying lens). At the bottom part of the IDE you should see values being printed. Put your hand close to the sensor and look at these values. Mind that if you blind the sensor you will not get the greatest value this sensor can output. Place your hand close, but not that close that you almost touch the sensor. The maximum value you see printed will be the maximum value of the sensor's range.

   Now that we know the range of our sensor, we can use it to control the amplitude of a Pyo oscillator. We will do this with both ways of communication between the Arduino and Python. We will start with pySerial.

5.3.3.1   Using pySerial

The Arduino code of Sketch 5.3 should stay almost intact, with the only change being the value passed to the `delay()` function. Change this value to 10. Script 5.3 shows the Python code. The Pyo bits of this code should already look familiar, so we will not discuss them here.

Script 5.3 Controlling the amplitude of a sine wave with a proximity sensor and pySerial.

```
 1 import serial
 2 from pyo import *
 3
 4 ser = serial.Serial('/dev/ttyACM1', 115200)
 5 s = Server().boot()
 6
 7 amp = SigTo(0)
 8 a = Sine(freq=440, mul=amp).out()
 9
10 min_thresh = 80
11 max_thresh = 600
12 sensor_range = max_thresh - min_thresh
13
14 def readval():
15     # check if there is data in the serial buffer
```

```
16      while ser.in_waiting:
17          # read bytes until a newline character '\n'
18          serbytes = ser.readline()
19          # strip b' from the beginning of the string
20          # and the ' at the end of it
21          serstr = str(serbytes)[2:-1]
22          # if there is more than just the return and mewline
23          if len(serstr) > 4:
24              # strip "\r\n" from the end
25              val = int(serstr[:-4])
26              if val > max_thresh: val = max_thresh
27              val -= min_thresh
28              if val < 0: val = 0
29              val /= sensor_range
30              amp.setValue(val)
31
32 def close():
33      pat.stop()
34      ser.close()
35
36 pat = Pattern(readval, time=.01).play()
37 s.gui(locals())
```

In lines 10 to 12 we set the range of our sensor. A maximum value of 600 is used, as the maximum value printed with Sketch 5.3 while testing was slightly above 600. It is also possible that the sensor never goes down to 0, so a minimum threshold is also set to 80. The total range of the sensor is set by the difference between the two thresholds.

The readval() function includes comments that explain what happens. When reading data from the serial line with pySerial, we check if there are any data in the serial buffer by checking the in_waiting value of the Serial() class of this module. This value returns the number of bytes in the receive buffer. If there are data in the buffer, we read until we get a newline character, by calling readline(). We see that using Arduino's println() is not helpful only when we read the printed values in Arduino's serial monitor, but also in Python, as we can use these special characters to separate incoming values.

Bytes received from the serial line with pySerial are returned in the form of the bytes Python data type. If we print the incoming values to the console intact, they will have the form b'450\r\n'. We want to strip the b and the first single quote at the beginning, as well as the last single quote, right after the newline character. To do this, we convert the bytes to a Python string with the str() class, and strip these characters by way of indexing. We can do this by setting two specific indexes separated by a colon. We want to strip two characters at the beginning of the string, so we set the value 2 as the first index, and after the colon we set -1. A negative index will start from the end of the string. [2:-1] tells Python that we want the string starting from the second index and ending one index before the end of the string. The square brackets must go outside of the parenthesis, because inside the parenthesis we have a bytes data type, but we use the square brackets to index a string, which is the result of str(serbytes).

Once we convert from bytes to string and strip the unnecessary characters, We check if there are values included apart from the carriage return and the newline characters, by checking the length of the string. Since the Arduino does not wait for Python to ask for new data, but keeps on sending data every 10ms, we might receive a string with these special characters only. If there are more characters in the string, we strip the carriage return and newline characters from it, by indexing the string from its beginning (no index at the left side of the colon) to four characters before its end. At the same time, we convert from string to int. Then we make sure our value is within the specified thresholds, we offset it to start from 0 instead of 80, we make sure we don't go below 0, and we divide by the total range, to get a value between 0 and 1. We set this value to the amp object of the `SigTo()` class and we are done.

We need to poll the serial buffer frequently, so we use a `Pattern()` to call the `readval()` function every 10ms. Mind that the `time` attribute of `Pattern()` should be the same as the value we set to the `delay()` function in the Arduino code, only in Python it is expressed in seconds, rather than ms. As soon as we start our script and the audio server, we can play with the sensor by moving our hand (or another obstacle) closer or farther from the sensor and hear the oscillator fade in and out.

The last function of our script is `close()` in line 32. In this function we first stop the `Pattern()` object, and then we close the serial port. If you want to keep your script alive, but you need to free up the port of the Arduino, you need to call the `close()` method of the `Serial()` class. Note that both our function and the method of the `Serial()` class have the same name, but that does not produce a name clash, since in the case of the `Serial()` class, we have to prepend the name of the `Serial()` object to the method call.

5.3.1.2   Using pyFirmata

Script 5.4 includes the code to get the sensor data with the Firmata firmware. Upload the Standard Firmata code to your Arduino board, and run the code in Script 5.4 to get the same result with Script 5.3.

Script 5.4 Controlling the amplitude of a sine wave with a proximity sensor and Firmata.

```
 1 from pyfirmata import Arduino, util
 2 from pyo import *
 3
 4 board = Arduino('/dev/ttyACM1')
 5 s = Server().boot()
 6
 7 # create an iterator thread to not constantly
 8 # send data to serial until it overflows
 9 it = util.Iterator(board)
10 it.start()
11 # initialize the pin you want to read
12 # 'a' means analog, 0 is the pin nr
13 # and 'i' means input
14 analog_pin = board.get_pin('a:0:i')
15
16 amp = SigTo(0)
17 a = Sine(freq=440, mul=amp).out()
18
19 min_thresh = 0.2
20 max_thresh = 0.65
21 sensor_range = max_thresh - min_thresh
22
23 def readval():
```

```
24     val = analog_pin.read()
25     if val > max_thresh: val = max_thresh
26     val -= min_thresh
27     if val < 0: val = 0
28     val /= sensor_range
29     amp.setValue(val)
30
31 def close():
32     pat.stop()
33     board.exit()
34
35 pat = Pattern(readval, time=.01).play()
36 s.gui(locals())
```

There are a few things we need to discuss here. From the documentation of pyFirmata, we read that we must create an `Iterator()` object, otherwise we will be sending data to the serial line constantly until it overflows. This happens in line 9, and in line 10 we start the iterator. In line 14 we initialise the analog pin 0 as an input, with the comment above it explaining what the string means.

In lines 19 to 21 we set our thresholds, only this time we use floats, because the Firmata firmware outputs analog values as floats in the range between 0 and 1. Reading the values of the analog pin is easier and more intuitive than it is in Script 5.3, as all we need to do is call the `read()` method of the `analog_pin` object. We then constrain the value to our thresholds, divide by our range to bring it back to the range between 0 and 1, and finally set the value to the `SigTo()` object. The last difference between the Firmata and the pySerial version is the `close()` function. In there, we call the `exit()` method of the `Arduino()` class, instead of a `close()` method, as the `Arduino()` class does not have any `close()` method, but only `exit()`.

5.3.2   *Using Two Proximity Sensors to Control the Amplitude and the Frequency of an Oscillator*

We can now move on and add one more sensor to control the frequency of the oscillator, and simulate a Theremin. Again, we will use both approaches to the communication between Python and the Arduino.

Ingredients:

- Python3
- The Pyo module
- The wxPython module, if you want to use Pyo's GUI
- The pySerial module
- The pyFirmata module
- A text editor (preferably an IDE with Python support)
- An Arduino Uno
- The Arduino IDE
- A breadboard
- Two infra-red proximity sensors
- A terminal window in case your editor does not launch Python scripts

Process:

Adding one more sensor does not mean we have to repeat the code we wrote once more. To make our code elegant, we will use loops to iterate over all of our sensors and get their values. Before we discuss the code, we have to first understand the breadboard, a necessary tool when building test circuits. Figure 5.4 illustrates a small breadboard with wires indicating how it is wired internally. This board enables us to create circuits by inserting electronic parts and jumper wires and connect them in any way we want.

We can see that the holes running along the blue and red line on the top and bottom part of the breadboard are connected internally the way they are connected in Figure 5.4 with the wires. By convention, ground connections are wired with black jumper wires, and voltage connections are wired with red jumper wires. On the breadboard, the ground connections are labelled with a blue line, and the voltage connections with a red line. In the inner part of the breadboard, the internal connections are vertical, the way the inner, smaller wire illustrates. The notch in the middle of the board splits it in two parts. This is helpful when we want to use a component with many legs that should not short between them.

Figure 5.5 illustrates the circuit for two infra-red proximity sensors. We need a breadboard because we have to connect both sensors to the 5V and ground pins of the Arduino. There are more than one ground pins on the Arduino, but only one 5V pin, so a breadboard is necessary here. We connect the power lines, ground and 5V, of the Arduino to the breadboard's ground and voltage rows, and we do the same with the two sensors. This way, both sensors are powered up. The signal wires of the sensors can connect straight to the Arduino.

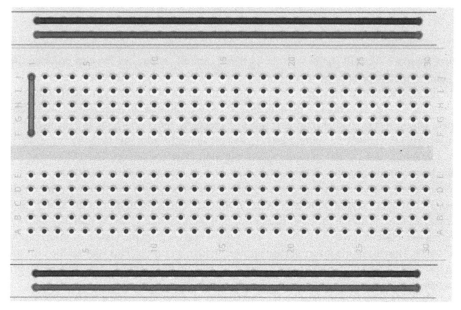

Figure 5.4 A breadboard with indications of its internal connections.

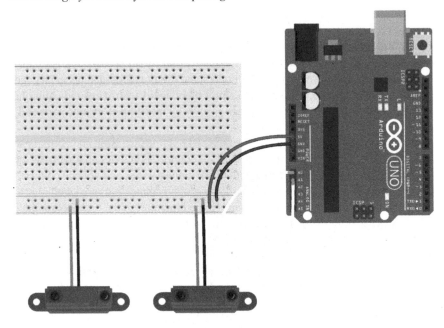

fritzing

Figure 5.5 Circuit with two infra-red proximity sensors connected to an Arduino Uno.

5.3.2.1   *Using pySerial*

The first approach we will see is with the pySerial module, where we have to program the Arduino ourselves. We can see the code in Sketch 5.4. In the `loop()` function we run a C++ for loop to iterate over the two sensors. This loop is a bit different that its Python version. We have to include the loop's elements in a parenthesis, where we initialise an integer with the value of 0, and then we test if it is smaller than 2. If this test succeeds, the loop's code is executed and then we return to the last part of the parenthesis, `i++`. This semantic adds one to the variable `i`. Once `i` has been incremented, we return to the test of the loop, where we check if `i` is less than 2. This way the loop will run as many times as the value we test against, in this case, two.

Inside the loop, we read the value of each of the two analog pins and then we print this information. To be able to separate the two values from each other, we first print the string `"sens"` for sensor, then we print the index of the sensor pin, then a white space, and finally the value of the analog pin. Note that we call `println()` only when we print the last bit of information, the sensor value. The rest of the information is printed with `print()`. This way all the information for each sensor will be unified with the carriage return and the newline characters being printed only when all the sensor information has been printed.

Sketch 5.4 Sending data for two proximity sensors.

```
1 void setup() {
2     Serial.begin(115200);
3 }
```

```
 4
 5 void loop() {
 6     for (int i = 0; i < 2; i++) {
 7         int sensVal = analogRead(i);
 8         Serial.print("sens");
 9         Serial.print(i);
10         Serial.print(" ");
11         Serial.println(sensVal);
12     }
13     delay(10);
14 }
```

The Python code to read the data is shown in Script 5.5. Since we have two sensors, we create a `SigTo()` object with two streams, by passing a list to its first argument. In our oscillator, we get each stream separately by indexing the same way we index lists in Python. In the `readvals()` function, once we strip the incoming bytes we test if the resulting string starts with "`sens`". If it does, we isolate the sensor index and its value. The string "`sens`" has four characters, so to isolate the sensor index, we write `serstr[4:5]`, so we can isolate the character after the fourth, and until the fifth. To isolate the sensor value, we get the string from the sixth character, because we have a white space after the sensor index, until four characters before the end. This process enables us to receives strings of the type "`sens0 145\r\n`" and isolate the data we want.

Once we isolate the information we need, we apply the thresholds. Since thresholding results in values in the range between 0 and 1, when the sensor index is 0, we multiply the value by 127 to get a range between 0 and 127. This is because in the `freq` attribute of the `SineLoop()` class, we pass an `MToF()` class with the Arduino value passed to its input. This class converts MIDI values to frequency. Since MIDI data are 7-bit, the maximum value we can get is 127, so we multiply the sensor value by this value and `MToF()` will do the conversion for us. In the last line of the `readvals()` function, we set the sensor value to the corresponding stream of the `SigTo()` object by using the isolated `index`.

Script 5.5 Controlling amplitude and frequency of an oscillator with pySerial.

```
 1 import serial
 2 from pyo import *
 3
 4 ser = serial.Serial('/dev/ttyACM1', 115200)
 5 s = Server().boot()
 6
 7 sigs = SigTo([0, 0])
 8 a = SineLoop(freq=MToF(sigs[0]),feedback=.1,mul=sigs[1]).out()
 9
10 min_thresh = 80
11 max_thresh = 600
12 sensor_range = max_thresh - min_thresh
13
14 def readvals():
15     while ser.in_waiting:
16         serbytes = ser.readline()
17         serstr = str(serbytes)[2:-1]
18         if serstr.startswith("sens"):
```

```
19                  index = int(serstr[4:5])
20                  # strip "sensX " from beginning, X is a number
21                  # and "\r\n" from the end
22                  val = int(serstr[6:-4])
23                  if val > max_thresh: val = max_thresh
24                  val -= min_thresh
25                  if val < 0: val = 0
26                  val /= sensor_range
27                  if index == 0:
28                      val *= 127
29                  sigs[index].setValue(val)
30
31 def close():
32     pat.stop()
33     ser.close()
34
35 pat = Pattern(readvals, time=.01).play()
36 s.gui(locals())
```

5.3.2.2   *Using pyFirmata*

Reading two sensors with pyFirmata is a bit simpler than the pySerial version. Upload the Standard Firmata sketch to your Arduino board and run the code in Script 5.6.

Script 5.6 Controlling amplitude and frequency of an oscillator with pyFirmata.

```
 1 from pyfirmata import Arduino, util
 2 from pyo import *
 3
 4 board = Arduino('/dev/ttyACM1')
 5 s = Server().boot()
 6
 7 it = util.Iterator(board)
 8 it.start()
 9
10 analog_pins = [board.get_pin('a:0:i'), board.get_pin('a:1:i')]
11
12 sigs = SigTo([0, 0])
13 a = SineLoop(freq=MToF(sigs[0]),feedback=.1,mul=sigs[1]).out()
14
15 min_thresh = 0.2
16 max_thresh = 0.65
17 sensor_range = max_thresh - min_thresh
18
19 def readvals():
20     for i in range(2):
21         val = analog_pins[i].read()
22         if val > max_thresh: val = max_thresh
23         val -= min_thresh
24         if val < 0: val = 0
25         val /= sensor_range
26         if i == 0:
27             val *= 127
```

```
28            sigs[i].setValue(val)
29
30 def close():
31      pat.stop()
32      board.exit()
33
34 pat = Pattern(readvals, time=.01).play()
35 s.gui(locals())
```

In line 10 we create a list with two `board.get_pin()` objects, one for each sensor. We pass the appropriate arguments to each, by setting the correct pin number in the argument string. In the `readvals()` function we run a loop where we read both sensors. To scale the range of the first sensor, we test against the `i` variable, as we don't need to isolate this index the way we did in the pySerial version. This is because with pyFirmata we explicitly ask for data from a given pin in the Python code, and the Arduino returns the requested information. With pySerial, the Arduino sends all the information without any requests from python, hence parsing the incoming data to isolate the index was necessary.

Documentation Page:

`MToF()`: https://belangeo.github.io/pyo/api/classes/utils.html#mtof

5.4 Controlling LEDs with Oscillators or by Detecting Attacks

We have already lit up the on-board LED of the Arduino by sending data from Python, but in this section we will see how we can use this feature to respond to audio we will process with Pyo.

5.4.1 Blinking LEDs by Detecting Attacks

We will first make an LED blink when we get an attack from incoming audio. We will use a microphone to receive external audio in Python. The microphone can either be your computer's internal microphone, or an external microphone, depending on your setup.

Ingredients:

- Python3
- The Pyo module
- The wxPython module, if you want to use Pyo's GUI
- The pySerial module
- The pyFirmata module
- A text editor (preferably an IDE with Python support)
- An Arduino Uno
- The Arduino IDE
- A breadboard
- An LED
- A 1 kOhm resistor
- A microphone
- A terminal window in case your editor does not launch Python scripts

Process:

We will be using the digital pin 3 instead of the on-board LED, so we will have to make a circuit for it. We could use the on-board LED, but by using another digital pin, we will learn how to build a circuit for an LED, since the on-board LED circuit is included on the Arduino board and we cannot see how it really works. Also, when we will control the LED with an oscillator, we will use Arduino's Pulse Width Modulation (PWM), a feature that is not available in digital pin 13. The tilde symbol next to pins 3, 5, 6, 9, 10, and 11, indicates that these pins have PWM capability.

Figure 5.6 shows the circuit. LEDs have two legs, a short one and a long one. The short one connects to ground, and the long one connects to the pin that will control the LED. A resistor must be placed somewhere along the circuit, be it between the Arduino's digital pin 3 and the long leg of the LED, or between the short leg of the LED and ground. If we omit this resistor, we will be creating a short circuit and the LED will be burnt.

Arduino's digital pin 13, in addition to including an on-board LED connected to it, also has an internal resistor, so we can connect an LEDs long leg to this pin and its short leg to the ground pin (which is next to pin 13), without using an external resistor. This holds though only for digital pin 13. All other digital pins need an external resistor. The resistor we use in this circuit is 1 kOhms. You can use a smaller resistor, down to 220 Ohms. The smaller the resistor, the brighter the LED will light up.

5.4.1.1   Using pySerial

We will first use the pySerial module. The Arduino code is shown in Sketch 5.5. When we receive the byte 1 from Python, we will light up the LED by setting its pin `HIGH`, and after 50ms we will switch it off by setting its pin `LOW`.

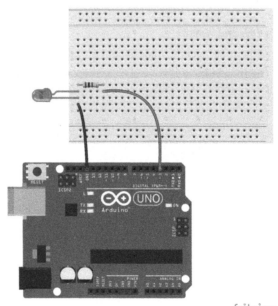

Figure 5.6 LED circuit with a 1 kOhm resistor.

Sketch 5.5 Arduino code for lighting up an LED when Pyo detects attacks.

```
1 void setup() {
2     pinMode(3, OUTPUT);
3     Serial.begin(115200);
4 }
5
6 void loop() {
7     if (Serial.available()) {
8         int val = Serial.read() - '0';
9         if (val == 1) {
10            digitalWrite(3, HIGH);
11            delay(50);
12            digitalWrite(3, LOW);
13        }
14    }
15 }
```

Script 5.7 shows the Python code to control this pin by detecting attacks from a microphone. In line 10 we get the input from the microphone with the `Input()` class. In line 11 we create an `AttackDetector()` and send the microphone's input to it, so it can detect the attacks from the incoming audio. When it detects an attack, it will output a trigger signal. In line 12 we create a `TrigFunc()` object. This class takes a PyoObject that sends trigger signals as its first argument, and a Python function as its second argument. When the PyoObject sends a trigger, `TrigFunc()` will call the function of its second argument. The function `Trig-Func()` calls, is the `lightup()` function that sends the byte 1 to the serial line. Run this script and make some percussive sounds close to the microphone, like clapping your hands, snapping your fingers, or hitting some object. The LED of your circuit should light up briefly and then turn off.

Script 5.7 Controlling an LED with pySerial by detecting attacks from a microphone.

```
1 import serial
2 from pyo import *
3
4 ser = serial.Serial('/dev/ttyACM1', 115200)
5 s = Server().boot()
6
7 def lightup():
8     ser.write(b'1')
9
10 mic = Input(0)
11 att = AttackDetector(mic)
12 tf = TrigFunc(att, lightup)
13
14 s.gui(locals())
```

5.4.1.2   Using pyFirmata

To replicate this setup with pyFirmata we must first upload the Standard Firmata sketch to our Arduino board. Then we must run the code in Script 5.8. The code should not need explanation as we have already seen all the classes and methods in previous scripts. We should note that the delay time between lighting up and turning off the LED is now included in the Python script by calling the `sleep()` function of the `time` module. Remember that this function expresses time in seconds, in contrast to Arduino's `delay()` function which expresses time in ms. We pass the value of 0.05 as the argument to `sleep()` to get a 50ms delay, like we did in the Arduino code in Sketch 5.5.

Script 5.8 Controlling an LED with pyFirmata by detecting attacks from a microphone.

```
 1 from pyfirmata import Arduino
 2 from time import sleep
 3 from pyo import *
 4
 5 board = Arduino("/dev/ttyACM1")
 6 s = Server().boot()
 7
 8 def lightup():
 9     board.digital[3].write(1)
10     sleep(.05)
11     board.digital[3].write(0)
12
13 mic = Input(0)
14 att = AttackDetector(mic)
15 tf = TrigFunc(att, lightup)
16
17 s.gui(locals())
```

Documentation Pages:

`AttackDetector()`: https://belangeo.github.io/pyo/api/classes/analysis.html#attackdetector
`TrigFunc()`: https://belangeo.github.io/pyo/api/classes/triggers.html#trigfunc

5.4.2   *Fading an LED In and Out with an Oscillator*

The next project will be to fade an LED in and out by sending the values of an oscillator to the Arduino. The circuit for this project is the same as the previous one, with an LED connected to digital pin 3 through a resistor.

Ingredients:

- Python3
- The Pyo module
- The wxPython module, if you want to use Pyo's GUI
- The pySerial module
- The pyFirmata module
- A text editor (preferably an IDE with Python support)
- An Arduino Uno
- The Arduino IDE
- A breadboard
- An LED

- A 1 kOhm resistor
- A terminal window in case your editor does not launch Python scripts

Process:

We will be using the output values of a sine wave oscillator to control the intensity of an LED. To do this, we need a digital pin with PWM capability. PWM works like a square wave oscillator. It has a fixed frequency, and we can control the percentage of its cycle that the pulse will be high, and the percentage that it will be low. If we set the PWM value half-way, for half of its cycle the square wave will output 5V, and for the other half it will output 0V. By setting a high frequency to this oscillation, we perceive this fast blinking as a dimly lit LED. When the PWM value is set lower, the intensity of the LED drops even more, and when the value goes higher, the intensity grows bigger. When the PWM value is at its full, the LED is fully lit, and does not go off at any point in the cycle of the square wave.

5.4.2.1   Using pySerial

As always, we will first use pySerial, so we have to write some Arduino code. This is shown in Sketch 5.6. We have some new elements in this sketch. In line 8 we define a static variable. This means that this variable will keep its value even after the function whose scope it belongs to exits, but it will still be available only within the scope of this function. The initial value we set to this variable is set only the first time we enter this function. After that, there is no initialisation of this variable any more, and it is not reset to 0 again.

Sketch 5.6 Arduino code to fade an LED in and out with PWM.

```
1 void setup() {
2      pinMode(3, OUTPUT);
3      Serial.begin(115200);
4 }
5
6 void loop() {
7      if (Serial.available()) {
8          static int val = 0;
9          byte in = Serial.read();
10         if (isDigit(in)) {
11             val = val * 10 + in - '0';
12         }
13         else if (in == 'v') {
14             analogWrite(3, val);
15             val = 0;
16         }
17     }
18 }
```

In line 10 we test if the incoming byte is a digit, and if it is, we execute line 11. In that line we assemble the incoming integer, digit by digit. If the incoming value is 126, the first digit that will arrive is 1, in ASCII, so 49. We multiply the variable val by 10, which yields 0, since the first time we run this function, the variable is initialised with a value of 0, and then we add the incoming ASCII value and subtract 48. This will yield 1. Then next digit will be 2. val is now

1, so multiplying by 10 yields 10. Adding 50 (ASCII of 2) and subtracting 48, yields 12. The next digit is 6. val is now 12 and multiplying by 10 yields 120. Adding 54 (ASCII of 6) and subtracting 48 yields 126, our initial value. It should be pointed out that in programming, the order of execution of calculations is the following: multiplications are first, divisions are next, additions follow, and subtractions are last. In line 11 we don't need to add any parentheses to define the order of execution, because we want the multiplication to be done first, then the addition, and the subtraction last (even though the last two calculations are inter-changeable in their order of execution). If we needed another execution order, then we would have to add parentheses to explicitly define the desired order.

To be able to tell whether the digits of our value have ended, we need to send a character that will denote this. In this sketch we choose the character 'v', since we are sending a value. When we receive this character, we write the stored value to the PWM pin with the analogWrite() function, and we reset the variable val. This last step is very important, because if we omit it, the variable will keep on shifting digits and adding values, going out of range in a few iterations. The analogWrite() function takes two arguments, the PWM pin to write to, and the value to write to that pin. Arduino's PWM range is 8-bit, with values from 0 to 255, 256 discrete steps.

The Python code is shown in Script 5.9. The only line that needs explanation is line 10. The Serial() class takes the bytes data type as its argument, to send bytes to the serial line. To get the oscillator value, we call its get() method that returns a float with the first output value of each sample block. Our oscillator has been scaled to a range between 0 and 1 in line 7, and we multiply it by 255, to get the full PWM range. We then make it an integer by enclosing it in the parenthesis of the int() class, and convert it to a string by enclosing it in the parenthesis of the str() class. We also add the string "v" that will denote the end of the digits of our value. The resulting string will be of the form "126v", and we pass this as the first argument to the bytes() class. The second argument of this class concerns the encoding type. Once we have converted the oscillator's output to a byte, we write it to the serial line.

Script 5.9 Fading an LED in and out with an oscillator and pySerial.

```
 1 import serial
 2 from pyo import *
 3
 4 ser = serial.Serial('/dev/ttyACM1', 115200)
 5 s = Server().boot()
 6
 7 sine = Sine(freq=.5, mul=.5, add=.5)
 8
 9 def lightup():
10     b = bytes(str(int(sine.get()*255))+"v", 'utf-8')
11     ser.write(b)
12
13 pat = Pattern(lightup, time=.01).play()
14
15 s.gui(locals())
```

5.4.2.2   Using pyFirmata

To get the same result with pyFirmata we have to upload the Standard Firmata sketch to our Arduino board, and run the code shown in Script 5.10. In line 9 we create a board.get_pin() object, and we pass the string "d:3:p" as its argument. "d" stands for "digital", 3 is the pin number, and "p" stands for "PWM". This way we tell the Arduino to set the digital pin 3 as a

PWM output. pyFirmata takes care of scaling the values, so we don't need to multiply the unipolar output of the oscillator by 255. The 0 to 1 range is scaled to the 0 to 255 range internally, so we can pass the value of the oscillator's output as is.

Script 5.10 Fading an LED in and out with an oscillator and pyFirmata.

```
1 from pyfirmata import Arduino
2 from pyo import *
3
4 board = Arduino("/dev/ttyACM1")
5 s = Server().boot()
6
7 sine = Sine(freq=.5, mul=.5, add=.5)
8
9 pwm = board.get_pin("d:3:p")
10
11 def lightup():
12     pwm.write(sine.get())
13
14 pat = Pattern(lightup, time=.01).play()
15
16 s.gui(locals())
```

5.5   Creating a Granular Synthesizer with Arduino and Python

We are now equipped with the necessary tools to build a complete application with hardware and software. We will revisit the Granular Synthesis Pyo classes from Chapter 3, and we will control them with potentiometers and tactile switches.

Ingredients:

- Python3
- The Pyo module
- The wxPython module, if you want to use Pyo's GUI
- The pySerial module
- The pyFirmata module
- A text editor (preferably an IDE with Python support)
- An Arduino Uno
- The Arduino IDE
- A breadboard
- Six 10 kOhm potentiometers
- Four tactile switches
- Four LEDs
- Four 10 kOhm resistors
- Four 1 kOhm resistors
- A terminal window in case your editor does not launch Python scripts

Process:

We will use all six analog inputs of the Arduino, and a few digital pins for both input and output. Figure 5.7 illustrates the circuit. This is the most advanced circuit we have built in this chapter, so it needs some explaining. On the left part we have six 10 kOhm rotary potentiometers. These adjust the voltage on their middle pin according to their position, as we twist them. One of their outer pins must connect to 5V, and the other to ground. The middle pin connects to an analog pin of the Arduino. We have already seen the LED circuit, so they should not need to be explained.

The tactile switches, also referred to as push-buttons, have their pins connected internally in pairs. The way they are shown in Figure 5.7, their pins are connected internally on each side, left and right. When the button is pressed, all pins connect to each other. One pin must connect to 5V, and the other must connect to a digital pin of the Arduino. With these connections only, when the button is not pressed, the corresponding digital pin will be left floating, as it will not be connected to anything, resulting in reading noise. To avoid this, we connect the button pin that connects to the Arduino, to ground. If we omit to use a resistor though, when we press the button, we will create a short circuit between 5V and ground. By connecting the pin that connects to the Arduino, to ground, through a 10 kOhm resistor, we avoid this short circuit when the button is pressed, and we also avoid having a floating pin reading noise when the button is not pressed.

This type of resistor is called a pull-down resistor. If we invert the connections and connect the pin that connects to the Arduino, to 5V, through the resistor, and the other button pin to ground, the resistor will be a pull-up resistor. With a pull-down resistor, when we press the button we read a high voltage, and when we release it, we read a low voltage. With a pull-up resistor the readings are inverted. The Arduino includes internal pull-up resistors for all its digital pins, but it is worth building the circuit explicitly to understand how it works.

One last thing to note concerns the breadboard. The one used in Figure 5.7 is longer than the one used in other figures in this chapter. These long breadboards separate their power rows to two, where in the middle of the board, the rows are disconnected. This is why there are two short

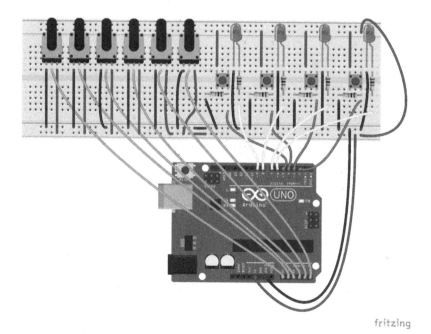

fritzing

Figure 5.7 Circuit for a granular synthesizer.

jumper wires at the bottom power rows, connecting the two sides, so power is transferred over to the side of the potentiometers.

5.5.1   Using pySerial

Let's start writing code for use of the pySerial module first. The Arduino code is shown in Sketch 5.7. The first new element we see in this code is comments in C++. These start with two forward slashes. Then we see two arrays in lines 2 and 3 that include the digital pins of the switches and the LEDs. In C++, we must declare the size of the array in square brackets when initialising it, and place its values, if known, in curly brackets. We use arrays for easier control and more flexible wiring.

Sketch 5.7 Arduino code for sending data for six potentiometers and four tactile switches, and receiving data to control four LEDs, to be used with pySerial.

```
 1  // create an array with the LED and switch pins
 2  int ledPins[4] = {6, 7, 8, 9};
 3  int switchPins[4] = {2, 3, 4, 5};
 4  // and an index for the array above
 5  int whichLed = 0;
 6
 7  void setup() {
 8      Serial.begin(115200);
 9
10      for (int i = 0; i < 4; i++) {
11          pinMode(switchPins[i], INPUT);
12          pinMode(ledPins[i], OUTPUT);
13      }
14  }
15
16  void loop() {
17      // read incoming data to control the LEDs
18      while (Serial.available()) {
19          static int val = 0;
20          byte in = Serial.read();
21          if (isDigit(in)) {
22              val = val * 10 + in - '0';
23          }
24          // a byte ending with 'l' denotes which LED to control
25          else if (in == 'l') {
26              whichLed = val;
27              val = 0;
28          }
29          // a byte ending with 'v' denotes an LED state
30          else if (in == 'v') {
31              digitalWrite(ledPins[whichLed], val);
32              val = 0;
33          }
34      }
35      // send the potentiometer values first
36      Serial.print("pots");
37      for (int i = 0; i < 6; i++) {
```

```
38              Serial.print(" ");
39              Serial.print(analogRead(i));
40         }
41         Serial.println();
42         // then send the switches
43         Serial.print("switches");
44         for (int i = 0; i < 4; i++) {
45              Serial.print(" ");
46              Serial.print(digitalRead(switchPins[i]));
47         }
48         Serial.println();
49         delay(10);
50  }
```

In the `setup()` function, we use a `for` loop to set the mode of the digital pins. This is the first place where we use these arrays. We could initialise the `i` variable of the loop for the switches with 2, and test if it is less than 6, so we could call `pinMode(i, INPUT)` without using an array. The way the wiring is done in Figure 5.7 though, the first switch in the loop is the right-most one. If we want the left-most switch to be the first, we can simply reverse the values of the `switchPins` array, and use the `for` loop the way it is used in the code, without needing to modify anything else. Using arrays also enables us to include the `pinMode()` calls for both input and output pins in the same loop.

In the `loop()` function, we first read incoming bytes in the serial buffer. This time we use a `while()` loop instead of an `if` test, so we can sort all the bytes in the buffer before we move on. The code of the `while()` loop is very similar to the one we used in the previous section, where we sent data from Python to the Arduino to fade an LED in and out. There is an additional step though, when we receive the '`l`' character. The bytes we will be sending from Python will be in the form "`2l1v`" where the first number will be the LED index, and the second number will be its state. The characters '`l`' and '`v`' enable us to distinguish what the previous numeric byte is supposed to be. When parsing this string, we will first assemble the numeric value, and when we reach a character, we will store this value where it has to be stored. By using arrays for the LED pins, we can index them starting from 0 – which is a standard practice in computer programming – instead of 6.

Once we are done reading the serial input, we print the values of the potentiometers and switches. This time, instead of sending each analog and digital value separately, we group the potentiometer values together, and we do the same for the switch values. We first print the string "`pots`" and then we run a loop to read and print the values in the six analog pins. We first print a white space to separate the values from each other, and to separate the first value from the "`pots`" string. Then we print the value of each analog pin. When the loop is done, we call `Serial.println()` to print the carriage return and the newline characters, so we can easily read this line that contains all the potentiometer values in Python. The same process is applied to the digital pins of the switches.

The Python code for this project using pySerial is shown in Script 5.11. This code contains many new concepts that we need to discuss.

Script 5.11 Python code for the granular synthesizer with pySerial.

```
1  import serial
2  from pyo import *
3
```

```
 4 ser = serial.Serial('/dev/ttyACM1', 115200)
 5 s = Server().boot()
 6
 7 potvals = [0, 0, 0, 0, 0, 0]
 8 switchvals = [0, 0, 0, 0]
 9 prev_switchvals = [0, 0, 0, 0]
10 prev_led = 0
11 prev_grains = 8
12 prev_basedur = 10
13
14 sigs = SigTo([0, 0, 0, 0, 0, 0])
15
16 snd_paths = ["./samples/audio_file1.wav",
17              "./samples/audio_file2.wav",
18              "./samples/audio_file3.wav",
19              "./samples/audio_file4.wav"]
20
21 snds = [SndTable(snd_paths[i]) for i in range(len(snd_paths))]
22
23 env = HannTable()
24 gran = Granulator(snds[0], env, pitch=sigs[0]/512,
25                   pos=(sigs[1]/1023)*snds[0].getSize(),
26                   dur=sigs[2]/1023, mul=sigs[5]/1023).out()
27
28 def init_leds():
29     # turn all LEDs off
30     for i in range(4):
31         b = bytes(str(i)+"10v", 'utf-8')
32         ser.write(b)
33     # light up first LED
34     b = bytes("011v", 'utf-8')
35     ser.write(b)
36
37
38 def readvals():
39     global prev_led, potvals, prev_grains, prev_basedur
40     while ser.in_waiting:
41         # read bytes until newline
42         serbytes = ser.readline()
43         # strip "b'" and "'" from bytes and convert to string
44         serstr = str(serbytes)[2:-1]
45         # read potentiometer values
46         if serstr.startswith("pots"):
47             # trim the "pots " part and split items
48             vals = serstr[5:-4].split()
49             try:
50                 # convert strings to integers
51                 potvals = [eval(i) for i in vals]
52                 #potvals = list(map(int, vals))
53                 sigs.setValue(potvals)
54                 # set attributes that don't take PyoObjects
55                 if int((potvals[3]/1023)*20+2) != prev_grains:
```

```
56                        prev_grains = int((potvals[3]/1023)*20+2)
57                        gran.setGrains(prev_grains)
58                    if int((potvals[4]/1023)*10+1)!= prev_basedur:
59                        prev_basedur = int((potvals[4]/1023)*10+1)
60                        gran.setBaseDur(prev_basedur/20)
61                except NameError:
62                    pass
63                except SyntaxError:
64                    pass
65
66            # read switch values
67            elif serstr.startswith("switches"):
68                # strip "switches " and split items
69                vals = serstr[9:-4].split()
70                # convert strings to integers
71                switchvals = [eval(i) for i in vals]
72                for i in range(len(switchvals)):
73                    # test against previous value of each switch
74                    if switchvals[i] != prev_switchvals[i]:
75                        # if value changed and switch is pressed
76                        if switchvals[i] == 1:
77                            # turn previous LED off
78                            b_str = str(prev_led)+"10v"
79                            # and current on
80                            b_str += str(i)+"11v"
81                            b = bytes(b_str, 'utf-8')
82                            ser.write(b)
83                            # update prev_led
84                            prev_led = i
85                            # set table to granulator
86                            gran.setTable(snds[i])
87                            # update pos to get right size
88                            gran.setPos((sigs[1]/1023)*\
89                                        snds[i].getSize())
90                        # update previous switch value
91                        prev_switchvals[i] = switchvals[i]
92
93 # light up the first LED
94 ca = CallAfter(init_leds, time=.5)
95
96 pat = Pattern(readvals, time=.01).play()
97
98 s.gui(locals())
```

The lists and variables in lines 7 to 12 are used to store values from the Arduino and to compare if a value has changed. The `potvals` list will store the values of the potentiometers, and the `switchvals` list will store the values of the switches. In line 9 we create a list to determine whether a switch value has changed. This is necessary as the baud rate of the serial communication is so fast, that even a momentary press on a push-button, will result in the Arduino sending the value 1 multiple times. We need to be able to determine when a switch change happens, so we can act on this change only once, and not for every single 1 that is sent from the Arduino. The rest of the variables starting with `prev_` are used for a similar reason.

In line 14 we create a `SigTo()` object with six streams, as many as the potentiometers in our circuit. These streams will control all the parameters of the `Granulator()` object we create in line 24. Note that the `grains` and `basedur` attributes of this class take an int and a float, respectively. They cannot be controlled by a PyoObject, so we will be using the respective values of the `potvals` list for these two. We could create the `SigTo()` object of line 14 with four streams instead, but then we would have to create two lists instead of the `potvals` list, and this approach would over-complicate things. The way the code is written, the `SigTo()` object has two streams too many, but this is very little overhead, compared to the complications we would get if we wanted to avoid it.

Our granular synthesizer will store four audio files, and we will be able to switch between them by using the push-buttons. In lines 16 to 19, we create a list with the paths to the sound files. Change the names according to your own sound files. These should be placed in a /samples subdirectory, in the directory of our script. In line 21 we create four `SndTable()` objects, one for each sound file. Then we create our Hanning window, and our granulator. Note that since we are using pySerial, the analog values will be in the 10-bit range, so in the kwargs of the `gran` object, we divide accordingly, to get the ranges we need. Since the `grains` and `basedur` attributes don't take PyoObjects, we omit them in the object initialisation. We will set values for these attributes later on.

In line 28 we define a function to light up the first LED. The LEDs will provide a visual feedback for the currently playing audio file. When the script is launched, the first audio file will be used, so we turn off all LEDs first, in case we re-run this script and another LED has remained lit, and then we turn on the first one. In line 38 we define the function that reads incoming data from the serial line. Since we want to test against variables that will hold previous values, we must declare them as global. Then we check if there are any bytes in the serial buffer.

In line 46 we test if the string "`pots`" is in the beginning of the received string. We do the necessary stripping and we split the string with the white space as a delimiter. Since we send all the potentiometer values in one go, separated by white spaces, line 48 will result in a list of strings, with the potentiometer values. We then `try` to convert these strings to integers by calling the built-in `eval()` function. The comment in line 52 provides another way to convert these strings to integers. We use the `try/except` mechanism, because we might read a string that contains other bytes, prior to "`pots`", in which case we will either get a `NameError`, or a `SyntaxError`. If this happens, it will most likely happen at the beginning. If you want to see this happening, remove the `try/except` mechanism, and use the code of the `try` block only. The program will not crash, you will only get the corresponding error in the traceback, and the program will continue. This mechanism is there only for the sake of elegance, so we don't get any traceback. Once the strings are converted, we pass the list with the integers to the `SigTo()` object.

In line 55 we determine whether we should change the number of grains. This attribute defaults to 8, and it should be an integer. We divide the potentiometer value by 1023, to get a value between 0 and 1, and we multiply by 20 and add 2, to bring it to a range between 2 and 22. We then convert it to an int by enclosing all these calculations to the parenthesis of `int()`. For the 10-bit resolution of the analog pins of the Arduino, a 22-value range is rather coarse, so we can safely test if the value we get from line 55 is different from the last value we stored for the number of grains. If it is, we update the `prev_grains` variable and call the `setGrains()` method of our granulator. We do the same for the `basedur` attribute in line 58, only the range is now between 1 and 11. This attribute though takes a float, so we call the `setBaseDur()` method with the `prev_basedur` variable divided by 20. Computers suffer from imprecision when it comes to floats, so it is much safer to test against an integer of a course range and determine if a value has changed, and then divide this integer to get the float you want.

In line 67 we read the values of the switches. Here no `try/except` mechanism is needed, because the switch values follow the potentiometer values. If an exception is raised, that will be caught in the `try/except` block above. Once we get the switch values as integers, we test if any of them has changed, and if it is being pressed, instead of being released. If this is the case, we turn off the currently lit LED and turn on the LED of the switch we are pressing. We then update the `prev_led` variable and change the `SndTable()` that our granulator is reading from. Since the reading position is dependent on the size of the table, we call the `setPos()` method explicitly, so we can multiply the `SigTo()` stream that controls it, by this size. PyoObjects are being updated at the audio rate, but getting the size of a table happens only when we explicitly call its `getSize()` method. If we omit this line, when we change a table, the multiplying coefficient will be still the size of the first audio sample.

Finally, in line 91 we update the current item of the `prev_switchvals` list. Pay attention to the indentation here. This update happens one level above the test of whether the switch is being pressed or not. Updating this list must be done whether we press or release the switch, otherwise, if we update it only when we press it, the list will stop being updated after the very first press, and the program will not work properly.

In line 94 we create a `CallAfter()` object to delay the initialisation of the LEDs a bit. If we call the `init_led()` function as soon as the script launches, the bytes sent will not go through and the LEDs will not change. Upload the Arduino sketch to your board and run this Python script. The sixth potentiometer will control the amplitude of the granular synthesizer, so you will need to turn this up. Experiment with this little instrument and you will most likely get interesting results. Note that the two potentiometers that control the `grains` and `basedur` attributes don't function at the audio rate. When you change any of these two values you will hear a slight click. It is also possible that a potentiometer is at a position that switches between two values in their coarse range. This is because the analog pins have some noise, and their values fluctuate a bit. If you hear constant clicks and get constant changes in the sound, twist the potentiometer that causes this a bit, and it should stop.

5.5.2 *Using pyFirmata*

Let's now see how we can create the same synthesizer with the Firmata firmware. Upload the Standard Firmata sketch to your Arduino board, and run the code in Script 5.12. Most of the concepts have been discussed in the pySerial version of this project.

Script 5.12 Python code for the granular synthesizer with pyFirmata.

```
 1 from pyfirmata import Arduino, util
 2 from pyo import *
 3
 4 board = Arduino('/dev/ttyACM1')
 5 s = Server().boot()
 6
 7 potvals = [0, 0, 0, 0, 0, 0]
 8 switchvals = [0, 0, 0, 0]
 9 prev_switchvals = [0, 0, 0, 0]
10 prev_led = 0
11 prev_grains = 8
12 prev_basedur = 10
```

```
13
14 sigs = SigTo([0, 0, 0, 0, 0, 0])
15
16 snd_paths = ["./samples/audio_file1.wav",
17               "./samples/audio_file2.wav",
18               "./samples/audio_file3.wav",
19               "./samples/audio_file4.wav"]
20
21 snds = [SndTable(snd_paths[i]) for i in range(len(snd_paths))]
22
23 it = util.Iterator(board)
24 it.start()
25 analog_pins = [board.get_pin('a:'+str(i)+':i')
26                for i in range(6)]
27 switch_pins = [board.get_pin('d:'+str(i+2)+':i')
28                for i in range(4)]
29 led_pins = [board.get_pin('d:'+str(i+6)+':o')
30             for i in range(4)]
31
32 env = HannTable()
33 gran = Granulator(snds[0], env, pitch=sigs[0]*2,
34                   pos=sigs[1]*snds[0].getSize(),
35                   dur=sigs[2], mul=sigs[5]).out()
36
37 # turn all LEDs off
38 for i in range(len(led_pins)):
39     led_pins[i].write(0)
40 # light up first LED
41 led_pins[0].write(1)
42
43
44 def readvals():
45     global prev_led, potvals, prev_grains, prev_basedur
46     for i in range(6):
47         potvals[i] = analog_pins[i].read()
48     sigs.setValue(potvals)
49     # set grains and basedur that don't take PyoObjects
50     if int(potvals[3]*20+2) != prev_grains:
51         prev_grains = int(potvals[3]*20+2)
52         gran.setGrains(prev_grains)
53     if int(potvals[4]*10+1) != prev_basedur:
54         prev_basedur = int(potvals[4]*10+1)
55         gran.setBaseDur(prev_basedur/20)
56     for i in range(4):
57         switchvals[i] = switch_pins[i].read()
58         if switchvals[i] != prev_switchvals[i]:
59             if switchvals[i] == 1:
60                 led_pins[i].write(1)
61                 led_pins[prev_led].write(0)
62                 prev_led = i
63                 # set table to granulator
64                 gran.setTable(snds[i])
```

```
65                        # update pos attribute to get right size
66                        gran.setPos(sigs[1]*snds[i].getSize())
67                    # update previous switch value
68                    prev_switchvals[i] = switchvals[i]
69
70
71 pat = Pattern(readvals, time=.01).play()
72
73 s.gui(locals())
```

In lines 25, 27, and 29, we create lists with `board.get_pin()` objects, using list comprehension, with the appropriate mode for each group, input or output. In lines 38 to 41, we turn off all LEDs and light up the first one. We can call this function immediately as with Firmata the serial connection has been settled before we reach this line in our program. The rest of the code is very similar to the pySerial version. Remember that with pyFirmata, the analog values are in the range between 0 and 1, so we don't need to divide by 1023, like we did with pySerial.

5.6   Pros and Cons of pySerial and pyFirmata

Even though pyFirmata provides an intuitive interface to communicate with the Arduino, plus we don't have to do any programming on the Arduino side, but we can do everything in Python, pySerial can prove to be more flexible in certain occasions. One such occasion is with the ultra-sonic range finder. A common mistake with proximity sensors is trying to use an ultra-sonic sensor with Firmata, instead of an infra-red. The latter sends an analog signal to the Arduino, and we can easily read this signal with pyFirmata's interface. The ultra-sonic sensor works in a similar way to the infra-red. This sensor sends an ultra-sonic pulse and receives the echo of it, as it bounces back when an obstacle is in front of it. Its difference to the infra-red sensor is that it sends digital data, as it does not include circuitry to convert the time difference to voltage. It is controlled by explicitly sending a trigger digital pulse, and reading the echo digital pulse back in the Arduino code. Then we calculate the distance based on the time difference in the Arduino code explicitly. Sketch 5.8 shows code to control an ultra-sonic sensor, copied from Arduino's Project Hub. If we try to use this sensor with Firmata, we will have to replicate the Arduino code in Python to get the correct results, but this means that we will have to replicate the `pulseIn()` function from line 17 as well. In this case, it is easier to use the code of Sketch 5.8 and get the data in Python with pySerial instead of using Firmata.

Sketch 5.8 Ultra-sonic proximity sensor Arduino code.

```
1 int distance, time;
2
3 void setup() {
4       pinMode(7,OUTPUT); //Trig Pin
5       pinMode(6,INPUT); //Echo Pin
6
7       Serial.begin(115200);
8 }
9
10 void loop() {
11      digitalWrite(7,LOW); //Trig Off
12      delayMicroseconds(20);
13      digitalWrite(7,HIGH); //Trig ON
```

```
14      delayMicroseconds(20);
15      digitalWrite(7,LOW); //Trig Off
16      delayMicroseconds(20);
17      time = pulseIn(6, HIGH);//TO RECEIVE REFLECTED SIGNAL
18
19      distance= time*0.0340/2;
20
21      Serial.print("dist: ");
22      Serial.println(distance);
23      delay(10);
24  }
```

Another occasion where pySerial can prove to be more flexible is when we want to read analog values with the ResponsiveAnalogRead Arduino library. This library is very effective in eliminating noise in the analog pins, without introducing too much latency. Usually, when we want to filter out the noise in the analog pins, we read a pin many times, accumulate these readings, and divide by the number of times we read, to get a mean value. This technique applies latency.

Another way to smooth the jitter from the analog pins of the Arduino is to use some kind of lowpass filter. This can be done in Python with Pyo's filters, like `ButLP()`, but again, the more we want to smooth out the analog values, the more the latency we will introduce. The ResponsiveAnalogRead library is introducing less latency than many other approaches. If we want to use this library, we will have to write our own Arduino code and read the data with pySerial. It is possible to integrate this library to the Firmata firmware, but this negates the scope of this firmware, which is to use one language for your project, and not having to program its various elements, each in a different language. If we want to integrate the ResponsiveAnalogRead library to Firmata, we will have to write Arduino code, and that is not what Firmata is about.

Besides the ultra-sonic sensor and the ResponsiveAnalogRead library, there are many more setups that would require translating code from Arduino to Python, to be used with pyFirmata. These setups include multiplexer chips that increase the number of analog pins, or shift registers, that increase the number of digital pins. Both chips come with their own Arduino code. Again, we could use Firmata and replace what happens in the Arduino code with Python, without even modifying the Standard Firmata sketch, but this is probably more of a hassle than a facilitation, plus it is likely that doing everything in Python will be slower than the equivalent Arduino code, combined with pySerial. Depending on the complexity of your project, you might find that pySerial is the preferred method over pyFirmata. Still, if a project is not very complicated in its electronics, pyFirmata can be very effective and provide code integrity, as everything will be done in one language only.

5.7 Conclusion

We have seen how we can connect Python to the real world through physical computing. We covered the communication between the Arduino and Python in both directions, and saw how we can get and send both digital and analog signals. The few projects that we realised in this chapter are enough to provide the necessary background for further experimentation and growth in project complexity. You are encouraged to take these projects, or your own, one step further and build a housing for the Arduino and the other electronic components. Providing guidance to hardware project building is beyond the scope of this book, but with a little bit of inspiration and some online searching, you could get quite far with such a project. You could even try to embed the computer that runs the Python scripts into your hardware, by using a Single Board Computer (SBC), like the Raspberry Pi, or an equivalent. Building digital synthesizers has never been easier, and if you find this appealing, it can be a very rewarding process.

5.8   Exercises

5.8.1   Exercise 1

Invert the readings of the sensor that controls the amplitude of the DIY Theremin in Script 5.5 or 5.6, so that the oscillator is muted when your hand is close to the sensor. This is actually how the Theremin works.

Tip: Use an `elif` test after the `if` test in line 27 or 26, in Script 5.5 or 5.6 respectively, where we test against the index of the sensor.

5.8.2   Exercise 2

Use a logarithmic curve to control the frequency of the DIY Theremin in Script 5.5 or 5.6, to have finer control over it.

Tip: You will have to raise the value of the sensor to a power of less than one. You don't need any Pyo classes, standard Python maths classes are sufficient.

5.8.3   Exercise 3

From Script 5.3 and for some scripts after that, we include a function that stops the `Pattern()` object and closes the serial port. Define a function that does the opposite, so you can reopen a port if you close it. Do this for pySerial.

Tip: Check the API of the `Serial()` class to determine which method to use. Mind the order of operations between `Serial()` and `Pattern()`.

Documentation Page:

`Serial()`: https://pyserial.readthedocs.io/en/latest/pyserial_api.html

5.8.4   Exercise 4

Light up one of two LEDs, depending on the amplitude of live audio input, where one LED will be lit when the input has low amplitude, and the other will be lit when the amplitude rises above a certain threshold.

Tip: Use the `Follower()` class, combined with `Compare()`, to determine which of the two LEDs should be lit.

Documentation Page:

`Follower()`: https://belangeo.github.io/pyo/api/classes/analysis.html#follower

Notes

1 A macro is a preprocessor definition of some value. In C++ macros are written in upper case letters by convention.
2 American Standard Code for Information Interchange, pronounced ASS-kee, is a character encoding standard.

Bibliography

Drymonitis, A. (2015) 'Introduction to Arduino', in *Digital Electronics for Musicians*. Berkeley, CA: Apress, pp. 51–96. Available at: https://doi.org/10.1007/978-1-4842-1583-8_2.

6   Communicating with Other Software with OSC

In this chapter we will look at how we can communicate with other software over a local network with the OpenSoundControl (OSC) protocol. We will experiment with software running on a different machine, like a smart phone, and software running on the same computer as the Python scripts we write.

What You Will Learn

After you have read this chapter you will:

- Know what OSC is
- Know how to connect your smartphone to Python
- Know how to connect your computer to other computers
- Know how to connect Python to other software and exchange messages
- Be able to utilise networked communication in a musical context

---

6.1   What is OSC?

OSC is a "protocol for communication among computers, synthesizers, and other multimedia devices" (Wright and Freed, 1997). It was developed by Matthew Wright and Adrian Freed, at the Center for New Music & Audio Technologies (CNMAT), and it was first presented at the International Computer Music Conference (ICMC) in 1997. As an alternative to MIDI, it overcomes flaws in the latter concerning transmission rates and numeric precision. Even though it initially aimed at providing a communication protocol that is more flexible than MIDI, OSC is often found alongside MIDI. It is often used as a network protocol, where different devices can communicate with each other over a local network, or different software in the same machine can send and receive messages between them. Initially targeting computer music software and hardware, OSC is now used by artists from various fields, even if no audio is included in their projects.

6.2   Available Python Modules for Sending and Receiving OSC Messages

Python includes many modules that support the OSC protocol, and Pyo itself includes classes for it. In the scripts we will write in this chapter, we will use the Pyo classes for OSC, but

DOI: 10.4324/9781003386964-6

we will find ourselves in need of another Python module that will facilitate some processes. This extra module is python-osc, and you can install it with Pip, by typing `pip install python-osc` in a terminal. Both this module and Pyo's classes, can send and receive various types of messages.

An advantage the python-osc module has over the Pyo OSC classes, is that we can receive data without knowing the receiving addresses in advance. This will be helpful when we try to connect a smart phone to a Python script, as the app we will be using does not specify its sending addresses. We will use this module to determine these addresses, and we will then use Pyo's classes to control our audio Python scripts.

6.3   Receiving and Sending OSC Messages between Your Smartphone and Pyo

One of the most popular smart phone apps for OSC is TouchOSC. This app is available for both Android and iOS, and you can get it from the respective app store. It is a low-cost commercial app that is often used by artists. If you want to use another app instead, you are welcome to do so. All the steps we will follow in Python, apply to any app used. The steps we will follow to set up the mobile app though will not apply to other apps, but they should give you an idea of how to set up yours.

The computer running the Python script, and the smart phone, will be communicating wirelessly. To achieve this, you need to have both machines connected to the same router. It is possible that your router has proxies that block OSC messages from being transmitted. In this case, you can create a hotspot on your mobile, and choose this network on your computer.

Ingredients:

- Python3
- The python-osc module
- The Pyo module
- The wxPython module, if you want to use Pyo's GUI
- A text editor (preferably an IDE with Python support)
- A smart phone, Android or iOS
- The TouchOSC mobile app
- A terminal window in case your editor does not launch Python scripts

Process:

In the TouchOSC app, you have to go to Settings and set the IP address of your computer, together with the outgoing port. To get your computer's IP, on macOS you have to go to System Preferences → Network, and click on the Wi-Fi connection that is active. The IP address will appear in the window. On Windows, go to Start → Settings → Network & Internet → Wi-Fi and select the wireless connection your computer is connect to. Under Properties, you can find your IP address, next to IPv4 address. On Linux, open a terminal window and type `hostname -I`. The IP of your computer will be printed on the terminal console. Type your computer's IP address to the "Host" field in TouchOSC's settings. The outgoing port can be anything you want, above 1024. For our purposes, we will use the port 9020. Choose the "Simple" layout in the Settings, and press "Done". Figure 6.1 illustrates the four different pages of this layout.

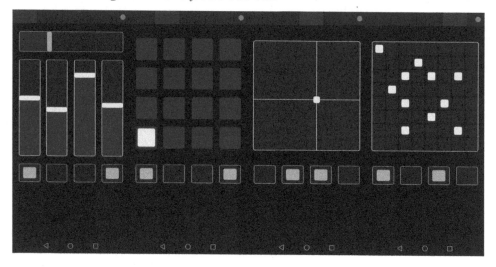

Figure 6.1 The four pages of the "Simple" layout in TouchOSC.

6.3.1 Receiving OSC Messages from Your Smartphone in Pyo

One caveat with TouchOSC is that we don't know the addresses each of these widgets sends OSC messages to. The OSC protocol supports highly customisable URL-style address naming, with symbolic names separated by forward slashes. TouchOSC prepends the page number in its addresses by default, and then uses the name of the widget concatenated to the widget index. In the first page, the left-most shown in Figure 6.1, the vertical sliders send messages to the following OSC addresses: /1/fader1, /1/fader2, /1/fader3, /1/fader4. The horizontal slider sends messages to the address /1/fader5. To find out the address of all the widgets in all four pages, we have to run the code in Script 6.1.

Script 6.1 Using the python-osc module to retrieve the OSC addresses of the TouchOSC widgets.

```
1   from pythonosc.dispatcher import Dispatcher
2   from pythonosc.osc_server import BlockingOSCUDPServer
3
4   def default_handler(address, *args):
5       print(f"{address}: {args}")
6
7   dispatcher = Dispatcher()
8
9   dispatcher.set_default_handler(default_handler)
10
11  ip = "192.168.43.187"
12  port = 9020
13
14  server = BlockingOSCUDPServer((ip, port), dispatcher)
15  server.serve_forever()
```

The IP address in line 11 will most likely be different on your machine. You should write the IP address of your computer, not your smart phone. When you run this script in a terminal window and start interacting with the widgets in the mobile app, you will see the address of the widget you are interacting with, being printed on the console, together with its value. To switch between pages in the TouchOSC app, touch the rectangles on the top part of the screen. The addresses and values printed to the console will be of the following type:

```
/1/fader5: (0.44370537996292114,)
/3/xy: (0.30381304025650024, 0.27750033140182495)
/4/multitoggle/6/3: (1.0,)
```

In the "Simple" layout, the only widget that sends two values is the XY pad in page 3. All the other widgets send one value only. The multitoggle widget is addressed with the column first and the row second. In the example above, the multitoggle pressed is the one on the sixth column and third row. Each page has four toggle switches at the bottom part of the screen, and page two has a four-by-four push-button matrix.

The addresses are received as a string, and the values as a tuple. We can therefore use standard Python indexing and conditional tests to isolate the addresses we want and their values. Interact with all the widgets of the layout, in all pages, to get all the OSC addresses. Once we know the addresses, we can use Pyo's OSC classes to receive data from the app. The code in Script 6.2 uses the `OscDataReceive()` class to receive messages from the app.

Script 6.2 Using Pyo's OscDataReceive() to receive data from TouchOSC.

```
1   from pyo import *
2
3   s = Server().boot()
4
5   def print_osc_data(address, *args):
6       if address == "/3/xy":
7           print(f"{address}: {args[0]}, {args[1]}")
8       else:
9           print(f"{address}: {args[0]}")
10
11
12  addresses = []
13  for i in range(5):
14      addresses.append("/1/fader"+str(i+1))
15
16  for i in range(4):
17      addresses.append("/1/toggle"+str(i+1))
18
19  for i in range(16):
20      addresses.append("/2/push"+str(i+1))
21
22  for i in range(4):
23      addresses.append("/2/toggle"+str(i+1))
24
25  addresses.append("/3/xy")
26
27  for i in range(4):
28      addresses.append("/3/toggle"+str(i+1))
29
```

```
30   for i in range(8):
31       for j in range(8):
32           addr_str = "/4/multitoggle/"+str(i+1)+"/"+str(j+1)
33           addresses.append(addr_str)
34
35   for i in range(4):
36       addresses.append("/4/toggle"+str(i+1))
37
38   osc_recv = OscDataReceive(9020, addresses, print_osc_data)
39
40   s.gui(locals())
```

The `OscDataReceive()` class can take as many addresses as we want in its second argument, whereas the first argument sets the port. In line 12 we initialise an empty list, and then we run a loop for each different widget group for each page, adding their addresses to the list. For the multitoggle widget, we run two nested loops, each with eight iterations, one loop for the columns and one for the rows. The XY pad widget has only one address, so we append it to the list without a loop.

The third argument of the `OscDataReceive()` class sets which Python function will be called when the object receives an OSC message. This function will be called with the receiving OSC address as its first argument, and a tuple with all the values, hence the asterisk in the second argument in the function definition in line 5. Since we know that only the XY pad widget sends two values, we test against the OSC address, and if it is "/3/xy" we print two values from the tuple, otherwise we print only one.

If you run Script 6.2 you will get a similar result to that of Script 6.1, only the values will now be printed isolated, instead of a tuple. This is convenient since this way we can easily set each value to any PyoObject we want. Once we run this script and establish a communication between the mobile app and Pyo, we can experiment with controlling sound with the app. Script 6.3 adds a `CrossFM()` class that has all of its attributes controlled by the five faders in page one.

Script 6.3 Control a CrossFM() class with TouchOSC.

```
1    from pyo import *
2
3    s = Server().boot()
4
5    sigs = SigTo([0, 0, 0, 0, 0])
6
7    def ctrl_xfm(address, *args):
8        if "/1/fader" in address:
9            index = int(address[-1:]) - 1
10           sigs[index].setValue(args[0])
11
12
13   addresses = []
14   for i in range(5):
15       addresses.append("/1/fader"+str(i+1))
16
17   for i in range(4):
18       addresses.append("/1/toggle"+str(i+1))
19
20   for i in range(16):
```

```
21        addresses.append("/2/push"+str(i+1))
22
23  for i in range(4):
24        addresses.append("/2/toggle"+str(i+1))
25
26  addresses.append("/3/xy")
27
28  for i in range(4):
29        addresses.append("/3/toggle"+str(i+1))
30
31  for i in range(8):
32        for j in range(8):
33            addr_str = "/4/multitoggle/"+str(i+1)+"/"+str(j+1)
34            addresses.append(addr_str)
35
36  for i in range(4):
37        addresses.append("/4/toggle"+str(i+1))
38
39  osc_recv = OscDataReceive(9020, addresses, ctrl_xfm)
40
41  sine = CrossFM(carrier=sigs[0]*200, ratio=sigs[1],
42                 ind1=sigs[2]*8, ind2=sigs[3]*8,
43                 mul=sigs[4]*0.5).mix(2).out()
44
45  s.gui(locals())
```

For this specific script, we could only add the OSC addresses we need, the five faders in page 1 in the TouchOSC app. All the addresses are appended to the list in case you want to experiment with more widgets. In line 5 we create a `SigTo()` object with five streams, and in the calling function of the `OscDataReceive()` object, we set the value to each stream separately by isolating the index of the OSC address.

Since the five faders are grouped under the address "/1/fader" with the fader index appended to that string, we can test against the common part of the string and isolate the last character of the string by indexing with a negative index in line 9, as we have already seen. We convert the string character to an integer, and we subtract one, because the OSC addresses start from 1. We use that value in line 10 to index the stream of the `SigTo()` object, so we can set the first and only tuple value to the appropriate stream.

In line 41 we create the `CrossFM()` object and pass each stream of the `SigTo()` object to its kwargs. Run this script and move the faders in page 1 to hear the result. Since we control the amplitude of the `CrossFM()` object with the fifth fader, you have to move the horizontal fader to produce sound.

6.3.1.1 *Another Way to Receive Floats via OSC in Pyo*

Pyo includes another way to receive OSC messages. Instead of calling a function whenever an OSC message comes in, we can create a PyoObject that outputs an audio stream with the float value that arrives through OSC. The class is `OscReceive()`. Note that this class works with float only, whereas `OscDataReceive()` works with any data type. The code in Script 6.3 can be rewritten to work with `OscReceive()`, and it is shown in Script 6.4.

Script 6.4 Control a CrossFM() class with TouchOSC and OscReceive().

```python
1   from pyo import *
2
3   s = Server().boot()
4
5   addresses = []
6   for i in range(5):
7       addresses.append("/1/fader"+str(i+1))
8
9   osc_recv = OscReceive(9020, addresses)
10
11  sine = CrossFM(carrier=osc_recv[addresses[0]]*200,
12                 ratio=osc_recv[addresses[1]],
13                 ind1=osc_recv[addresses[2]]*8,
14                 ind2=osc_recv[addresses[3]]*8,
15                 mul=osc_recv[addresses[4]]*0.5).mix(2).out()
16
17  s.gui(locals())
```

This version is simpler since we don't have to define any function and check the address of the message. We are creating five streams of the `OscReceive()` object since we pass a list with five items in its address argument. Note that we access the separate streams by way of indexing the `OscReceive()` object. The indexing though works like a Python dictionary, where the address string is used as the key. Since the first OSC address is "/1/fader1", the first stream is accessed with `osc_recv["/1/fader1"]`. To facilitate things, we access the separate streams by indexing the addresses list instead, since our OSC addresses are stored there. When dealing with floats only, this method is probably more suitable than the `OscDataReceive()` class. Be aware though that it doesn't seem to work with widgets that send more than one value, like the XY pad.

Documentation Pages:

`OscDataReceive()`: https://belangeo.github.io/pyo/api/classes/opensndctrl.html#oscdata receive

`OscReceive()`: https://belangeo.github.io/pyo/api/classes/opensndctrl.html#oscreceive

6.3.2 *Sending OSC Messages from Pyo to Your Smartphone*

Now that we have received messages from the smartphone in Pyo and have used these messages to control audio, let's send OSC messages from Pyo to the smartphone. When using TouchOSC, it can be useful to send data from Pyo to visualise certain aspects, like the amplitude of an oscillator or some other attribute for which we need to have visual feedback. We will use the same layout as we did for receiving messages from the smartphone, and the same page we used in Scripts 6.3 and 6.4. You are welcome to change the code and use it with any layout of the app, or any other app you like.

TouchOSC conveniently uses the same address for sending and receiving OSC messages. Since we know the addresses of the faders in page 1, we can use them to send data from Pyo with the `OscDataSend()` class. Establish a connection between your phone and your computer and run the code in Script 6.5. We will be sending the values of a sine wave to the first fader in page 1, to visualise the sine movement. The `OscDataSend()` class takes four arguments: the message type, the port, the OSC address, and the IP of the device that we will be sending messages to.

Note that if you create a hotspot on your mobile and connect your computer to that network, in TouchOSC the phone's IP address might appear as 0.0.0.0. In a local network though, the IP addresses must have the first three fields identical between all devices, and only the last field must be different. Since we create a network from our phone, the phone acts as a router, so the last field will be 1, like your home router most likely has the IP address 192.168.1.1. Compare the IP address in Script 6.1 with the one in Script 6.5 and you will see that this IP address principle is present.

Script 6.5 Sending OSC messages from Pyo to TouchOSC.

```
1   from pyo import *
2
3   s = Server().boot()
4
5   sine = Sine(freq=.5, mul=.5, add=.5)
6
7   osc_send=OscDataSend("f",9030,"/1/fader1",host="192.168.43.1")
8
9   def sendmsg():
10      osc_send.send([sine.get()])
11
12
13  pat = Pattern(sendmsg, time=.01).play()
14
15  s.gui(locals())
```

The data type kwarg of `OscDataSend()` is set as a string denoting the type. An "f" denotes a float, which is what we will be sending. In the `sendmsg()` function though we send this value inside a list. The `send()` method of the `OscDataSend()` class must take a list, even if we send one value only. Set the incoming port in the TouchOSC app to 9030 and launch the "Simple" layout. Run the Python script and look at the first fader. It should move up and down, in a sine fashion.

6.3.2.1 *Another Way to Send Floats via OSC in Pyo*

As with the `OscReceive()` class, Pyo includes its sending equivalent, `OscSend()`. Again, this class works with floats only, but it makes the code simpler. Functioning at the audio rate, this class sends an OSC message after each sample block, in contrast to `OscDataSend()`, which we have to ping explicitly, and we do this with `Pattern()`. The code in Script 6.5 is rewritten in Script 6.6 to work with `OscSend()`. As with `OscReceive()`, we don't need to define any function for `OscSend()`. We also don't need to strip the first sample of a sample block with `get()`, like we did in Script 6.5, in line 10. We just pass the `sine` object to the first argument of `OscSend()`, and the latter takes care of sending the first sample of each sample block.

Script 6.6 Sending OSC messages from Pyo to TouchOSC with OscSend().

```
1   from pyo import *
2
3   s = Server().boot()
4
5   sine = Sine(freq=.5, mul=.5, add=.5)
6
7   osc_send = OscSend(sine, port=9030, address="/1/fader1",
8                      host="192.168.43.1")
9
10  s.gui(locals())
```

Documentation Pages:

OscDataSend(): https://belangeo.github.io/pyo/api/classes/opensndctrl.html#oscdatasend
OscSend(): https://belangeo.github.io/pyo/api/classes/opensndctrl.html#oscsend

6.4   Sending and Receiving Messages between Pyo and an openFrameworks Program

In this last section, we are going to combine Pyo with an openFrameworks (OF) app. OF is a C++ toolkit for creative coding, mainly aiming at visuals. It is free and open source, and works on all three major OSes. You can download it from its official website,[1] and follow the instructions to install it on your system, depending on your OS. Providing detailed instructions for installing it, is beyond the scope of this chapter, but the instructions on the official website should be enough. If you want to follow this project, install OF first. We will write a bit of OF code, but we will not dive deep in explaining it. We have already seen C++ code in Chapter 5, when we programmed our Arduino boards, so the code in this chapter should look familiar.

Ingredients:

- Python3
- The Pyo module
- The wxPython module, if you want to use Pyo's GUI
- A text editor (preferably an IDE with Python support)
- The openFrameworks toolkit
- A terminal window in case your editor does not launch Python scripts

Process:

We will revisit the drum machine code from Chapter 3, in Script 3.28. We will build an OF app to control the triggering steps of each of the three drum samples. OF will send OSC messages to Python, so we can update the triggering lists, and Pyo will send OSC messages to OF, to highlight the current step. Similar to the Arduino, OF calls one function when an app is launched, and then runs a loop. The function called when launched is named setup(). After this function is called, OF calls two functions in a loop, one after the other. These functions are `update()` and `draw()`. Besides these three functions, there are some more functions that deal with keyboard input, mouse activity, the app window being resized, and others. In our project we will use the `mousePressed()` function to set the state of a step of any one of the three drum samples.

OF follows the standard C++ program structure, with two files comprising the program, a header file and the file with the implementation of the code. The header file contains the main OF class definition. This file is called ofApp.h. The other file, called ofApp.cpp, includes the code for all functions of the OF class. In C++, both files must share the same name, with only the suffix being different. OF also contains another file, main.cpp. This file sets the dimensions of the window of the app, and calls the main OF function that starts and runs the program we will write. Create a new project, preferably using the Project Generator.

If you use the Project Generator, give your project any name you like (avoid white spaces), don't change the "Project path:", and click the "Addons:" field to choose the ofxOsc addon. If you are not using the Project Generator, create a new directory in OF/apps/myApps (OF is the directory of your OF installation) with the name you want to give to your project, and copy everything from the emptyExample directory into your project directory. Change the name of any possible files named emptyExample.* (the asterisk denotes everything), to the name of your directory, keeping the suffixes intact. You should also add the following line to the addons.make file, if you are not using the Project Generator:

```
ofxOsc
```

The following steps are the same whether you use the Project Generator or not. Go to the src/ directory and copy the code in the OFcode chunks below, to the respective files. The main.cpp file, shown in OFcode 6.1, is almost identical to the original one, we just have to change the second argument to the `ofSetupOpenGL()` function, which is the height of the window of our app. The rest of the code in this file should remain the same.

OFcode 6.1: main.cpp
```
 1   #include "ofMain.h"
 2   #include "ofApp.h"
 3
 4   //=============================================================
 5  ·int main( ){
 6
 7       ofSetupOpenGL(1024,192, OF_WINDOW); // setup the GL context
 8
 9       // this kicks off the running of my app
10       // can be OF_WINDOW or OF_FULLSCREEN
11       // pass in width and height too:
12       ofRunApp( new ofApp());
13
14   }
```

In the ofApp.h file shown in OFcode 6.2, we insert four macros at the beginning of the file with the `#define` preprocessor directive. Note that the HOST macro is set to `"localhost"` since we will exchange OSC messages between software in the same computer. We add two OSC objects in lines 29 and 30, one for sending and one for receiving. Then we add some variables. The `metroState` variable is redundant in the version we will build here, but it will be necessary in one of the exercises of this chapter. In line 34 we initialise a two-dimensional array named `toggles`. This array will hold the states for each step of each drum sample, hence its first dimension is 3, the number of our samples, and the second dimension is 16, the number of the steps of the sequencer.

OFcode 6.2: ofApp.h

```
1    #pragma once
2
3    #include "ofMain.h"
4    #include "ofxOsc.h"
5
6    #define OFFSET 2
7    #define HOST "localhost"
8    #define OUTPORT 12345
9    #define INPORT 9030
10
11   class ofApp : public ofBaseApp{
12       public:
13           void setup();
14           void update();
15           void draw();
16
17           void keyPressed(int key);
18           void keyReleased(int key);
19           void mouseMoved(int x, int y);
20           void mouseDragged(int x, int y, int button);
21           void mousePressed(int x, int y, int button);
22           void mouseReleased(int x, int y, int button);
23           void mouseEntered(int x, int y);
24           void mouseExited(int x, int y);
25           void windowResized(int w, int h);
26           void dragEvent(ofDragInfo dragInfo);
27           void gotMessage(ofMessage msg);
28
29           ofxOscSender sender;
30           ofxOscReceiver receiver;
31
32           float rectSize;
33
34           int toggles[3][16];
35           int step;
36           int metroState;
37   };
```

The empty functions in the ofApp.cpp file, shown in OFcode 6.3, are standard functions of the OF main class. Since these functions are defined in the header file, we have to include them here too, even if they are empty. Explaining the whole code of ofApp.cpp is beyond the scope of this book. There are many online sources for learning OF, with the online ofBook[2] being a free source, generously provided by the developer community of OF.

OFcode 6.3: ofApp.cpp

```cpp
1   #include "ofApp.h"
2
3   //-----------------------------------------------------------
4   void ofApp::setup(){
5       ofBackground(0);
6
7       sender.setup(HOST, OUTPORT);
8       receiver.setup(INPORT);
9
10      rectSize = ofGetWindowWidth() / 16;
11
12      step = 0;
13      metroState = 1;
14
15      for (int i = 0; i < 3; i++) {
16          for (int j = 0; j < 16; j++) {
17              toggles[i][j] = 0;
18          }
19      }
20
21      ofSetLineWidth(2);
22  }
23
24  //-----------------------------------------------------------
25  void ofApp::update(){
26      while(receiver.hasWaitingMessages()) {
27          // get the next message
28          ofxOscMessage m;
29          receiver.getNextMessage(m);
30
31          if(m.getAddress() == "/seq") {
32              step = m.getArgAsInt32(0);
33          }
34      }
35  }
36
37  //-----------------------------------------------------------
38  void ofApp::draw(){
39      for (int i = 0; i < 3; i++) {
40          for (int j = 0; j < 16; j++) {
41              float x = j * rectSize;
42              float y = i * rectSize;
43              if ((j == step) && metroState) {
44                  ofSetColor(255, 0, 0);
45              }
46              else {
47                  ofSetColor(200);
48              }
49              ofDrawRectangle(x+(OFFSET/2), y+(OFFSET/2),
50                              rectSize-OFFSET, rectSize-OFFSET);
```

```
51                        if (toggles[i][j]) {
52                          ofSetColor(0);
53                          ofDrawLine(x+(OFFSET/2), y+(OFFSET/2),
54                                      x+(rectSize-(OFFSET/2)),
55                                      y+(rectSize-(OFFSET/2)));
56                          ofDrawLine(x+(rectSize-(OFFSET/2)), y+(OFFSET/2),
57                                      x+(OFFSET/2), y+(rectSize-(OFFSET/2)));
58                          ofSetColor(200);
59                        }
60                    }
61            }
62   }
63
64   //-----------------------------------------------------------
65   void ofApp::keyPressed(int key){
66
67   }
68
69   //-----------------------------------------------------------
70   void ofApp::keyReleased(int key){
71
72   }
73
74   //-----------------------------------------------------------
75   void ofApp::mouseMoved(int x, int y){
76
77   }
78
79   //-----------------------------------------------------------
80   void ofApp::mouseDragged(int x, int y, int button){
81
82   }
83
84   //-----------------------------------------------------------
85   void ofApp::mousePressed(int x, int y, int button){
86       ofxOscMessage m;
87       int xIndex;
88       int yIndex;
89       switch (button) {
90           case 0:
91               xIndex = x / rectSize;
92               yIndex = y / rectSize;
93               toggles[yIndex][xIndex] = !toggles[yIndex][xIndex];
94               m.setAddress("/seq");
95               m.addIntArg(yIndex);
96               m.addIntArg(xIndex);
97               m.addIntArg(toggles[yIndex][xIndex]);
98               sender.sendMessage(m, false);
99               break;
100          case 2:
101              metroState = !metroState;
102              m.setAddress("/metro");
103              m.addIntArg(metroState);
```

```
104             sender.sendMessage(m, false);
105             break;
106         case 1:
107             m.setAddress("/reset");
108             m.addIntArg(1);
109             sender.sendMessage(m, false);
110             break;
111         default:
112             break;
113     }
114 }
115
116 //----------------------------------------------------------
117 void ofApp::mouseReleased(int x, int y, int button){
118
119 }
120
121 //----------------------------------------------------------
122 void ofApp::mouseEntered(int x, int y){
123
124 }
125
126 //----------------------------------------------------------
127 void ofApp::mouseExited(int x, int y){
128
129 }
130
131 //----------------------------------------------------------
132 void ofApp::windowResized(int w, int h){
133
134 }
135
136 //----------------------------------------------------------
137 void ofApp::gotMessage(ofMessage msg){
138
139 }
140
141 //----------------------------------------------------------
142 void ofApp::dragEvent(ofDragInfo dragInfo){
143
144 }
```

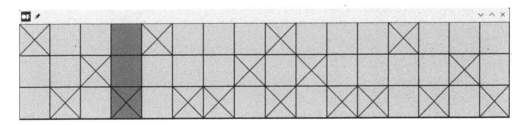

Figure 6.2 The OF app for the drum machine Python code.

In contrast to Python, C++ must be compiled to get an executable program. Compile this project with the appropriate method, depending on your OS, and then run the program. Figure 6.2 shows the OF app. When you start it, all boxes will be empty. By clicking inside a box, it gets activated. The boxes that are red are the ones of the current step of the sequencer. When you launch the app, the first three vertical boxes will be red. If you launch this app without running the Python code for this project, you will not get much out of it, so be patient and launch the Python script from Script 6.7 first.

Script 6.7 The drum machine Python code that communicates with the OF app.

```
 1  from pyo import *
 2
 3  s = Server().boot()
 4
 5  sample_paths = ["./samples/kick-drum.wav",
 6                  "./samples/snare.wav",
 7                  "./samples/hi-hat.wav"]
 8  # create a three-stream SndTable, and read only channel 0
 9  snd_tabs = SndTable(sample_paths, chnl=0)
10  # get a list of the durations
11  durs = snd_tabs.getDur(all=True)
12
13  # create a single list for all three samples
14  all_lists = [[0 for i in range(16)] for i in range(3)]
15
16  # get ms from BPM based on meter
17  metro = Metro(beatToDur(1/4, 120))
18  mask = Iter(metro, all_lists)
19  # create a dummy arithmetic object
20  beat = metro * mask
21
22  # create a three-stream TrigEnv()
23  player = TrigEnv(beat, snd_tabs, durs, mul=.5)
24  mix = Mix(player.mix(1), 2).out()
25
26  def new_list(address, *args):
27      all_lists[args[0]][args[1]] = args[2]
28      mask.setChoice(all_lists)
29
30
31  counter = Counter(metro, max=16)
32
33  osc_recv = OscDataReceive(12345, "/seq", new_list)
34  osc_send = OscSend(counter, port=9030,
35                     address="/seq",host="127.0.0.1")
36
37  metro.play()
38
39  s.gui(locals())
```

There are a few differences between Script 6.7 and Script 3.28 from Chapter 3. In Script 6.7, we don't initialise any active steps for any of the samples, since we will be doing this with the OF app. Since all steps will have a value of 0, we use a nested list comprehension in line 14, where the inner list has 16 items with the value 0, and the outer list has three items, which are all lists. In line 24 we create a `Mix()` object so we can get the audio in both speakers of a stereo setup. This is the solution to Exercise 7 of Chapter 3.

In line 31 we create a `Counter()` so we can keep track of the sequence count, and send it to the OF app via OSC. In lines 33 and 34 we create our OSC objects. For receiving messages we use the `OscDataReceive()` class because we will be receiving lists, and we can't handle lists with `OscReceive()`. These lists will contain the sample index, the step index, and the state value for that step, 1 for active and 0 for inactive. You can see that in the `new_list()` function.

For sending OSC message, we can use `OscSend()` instead of `OscDataSend()`, as we will be sending a single value, the step number. Note the IP address passed to the `host` kwarg. This is the local IP of a machine. It is equivalent to the "`localhost`" string in the OF app. For the code to run, you will have to copy the entire samples/ directory with the drum samples from Chapter 3, and paste it in the directory of the current Python script.

Run both Script 6.7 and the OF app, and you should see the red part shown in Figure 6.2 move from left to right. Click on some boxes and you should start hearing the respective sample being triggered when the step falls on a triggering box of that sample. The top row is the kick drum, the middle is the snare, and the bottom is the hi-hat.

Note that the two applications, OF and Python, can run on different machines. In that case, each sending OSC object should get the IP address of the other machine. A local network can be created through your home router, through a hotspot, or even through a straight connection with an Ethernet cable, if your computer has an Ethernet port. Without a router or hotspot, the IPs should be static and set manually. As already mentioned, the first three fields of the two IPs must be the same, and the last must be different between the two machines. The values for each field are 8-bit, so in the range between 0 and 255 inclusive. A usual IP is 192.168.x.y or 169.254.x.y, but it can really be anything within the 8-bit range, except from 255.255.255.x or any IP starting with 255. If you have your Wi-Fi on and you want to create a static IP for your Ethernet port, the IP 192.168.1.x is probably not a good idea, as your router is very likely to be providing such IP addresses. You can set something like 192.168.100.x, or an IP where the third field is different than that of your Wi-Fi IP.

6.5 Conclusion

We have seen how we can communicate with other software and hardware with the OSC protocol. There is a lot of potential in using OSC with music or other software as it can connect and sync to visuals, it can be controlled by custom controllers, or we can even create a networked system for electronic music. Networked performances are often the focus in various artistic events and conferences, and there is a lot of research going on in this field. Pyo provides all the necessary tools to support this feature in a flexible and intuitive way.

6.6 Exercises

6.6.1 *Exercise 1*

Use live input and control a TouchOSC widget as a VU meter, with the amplitude of this input.

6.6.2   *Exercise 2*

Use the XY pad of the TouchOSC app to control the carrier and modulator frequencies of an FM synth with a single widget.

Tip: Don't use the `FM()` class, because you cannot control the modulator frequency directly. Explicitly apply FM with two `Sine()` classes, as we did in Chapter 3.

6.6.3   *Exercise 3*

Use the "`/metro`" and "`/reset`" OSC addresses set in lines 102 and 107 in the ofApp. cpp file in OFcode 6.3. The value sent with the "`/metro`" address should start and stop the `Metro()` object, in Script 6.7, and the value sent with the "`/reset`" address should reset the sequencer in this script.

Tip: Remove line 37 from Script 6.7, so that `metro` doesn't start when the Python script is launched. Use a list with the three OSC addresses for the `address` kwarg of `OscDataReceive()`. In the `new_list()` function (which you might want to re-name, as it will do more than accept lists from OF, but make sure you set the same name in the function attribute of `OscDataReceive()`), set conditional tests for the incoming OSC addresses to determine where you will send the incoming values. Set another conditional test when controlling `metro`, based on the incoming OSC datum, to determine whether you will call its `play()` or `stop()` method. Finally, when you receive data with the "`/reset`" address, make sure you reset both `mask` and `counter`. The value sent with this address is insignificant. In the ofApp.cpp file, in line 13, initialise the `metroState` variable to 0. When you run this script, a right click of the mouse should start and stop the sequencer, and a middle click should reset it.

6.6.4   *Exercise 4*

Modify the drum machine Python code so that the OF app is launched from there and not manually.

Tip: You need to import the Popen class from the subprocess module. Then create a `Popen()` object and pass a list with a single item to its argument. Look up its documentation to determine how to use it.[3] The executable to the OF app is located in OF/apps/ myApps/nameOfApp/bin, where OF is the directory of your OF installation.

Notes

1 https://openframeworks.cc/
2 https://openframeworks.cc/ofBook/chapters/foreword.html
3 https://docs.python.org/3/library/subprocess.html#subprocess.Popen

Bibliography

Wright, M. and Freed, A. (1997) 'Open SoundControl: A New Protocol for Communicating with Sound Synthesizers', in *Proceedings of the 1997 International Computer Music Conference, ICMC 1997, Thessaloniki, Greece, September 25–30, 1997*. Michigan Publishing. Available at: http://hdl.handle. net/2027/spo.bbp2372.1997.033.

7   Machine Learning in Music

In this chapter we will learn the basics of Machine Learning (ML) and Neural Networks (NN) and how we can use them in a musical context. We will learn the underlying structure of an NN, and we will briefly go through some Python modules for ML and NNs. We will then learn how we can create NNs for our use cases and how to get meaningful data out of them, to use them musically.

What You Will Learn

After you have read this chapter you will:

- Be acquainted with ML and NN concepts
- Know the inner structure of an NN
- Be able to get meaningful data from an NN
- Be able to use NNs in a musical context

7.1   What are Machine Learning and Neural Networks

ML is a scientific field where an algorithm, running in computer code, tries to "learn" certain patterns from incoming data, so it can provide correct predictions when it is confronted with data it has never seen before. NNs are a subset of ML (Kinsley and Kukieła, 2020, p. 11). Even though they are "patterned after the interconnections between neurons found in biological systems" – which systems consist of billions of interconnected neurons – exchanging electric signals, they bear little resemblance to such systems (Priddy and Keller, 2005, p. 1).

An NN is a set of artificial neurons that interconnect and exchange numeric data. NNs are actually called Artificial Neural Networks (ANN), but we often omit the word "Artificial", and refer to this mechanism simply as Neural Networks. They were initially conceived in the 1940s but were not given much attention before 2010 (Kinsley and Kukieła, 2020, p. 12). A subset of NN is Deep Learning. NNs with two or more hidden layers fall in this category (Kinsley and Kukieła, 2020, p. 11). We will use Multi Layer Perceptrons (MLP) (Gardner and Dorling, 1998) in our projects, which, depending on their structure, can also be called Deep Neural Networks (DNN). These terms will be used throughout this chapter inter-changeably.

DOI: 10.4324/9781003386964-7

7.1.1   Applications of ML

ML is being included in a multitude of applications like language translation, image recognition, speech recognition, sentiment analysis of text, text generation, products suggestion, and others. We encounter ML in our everyday life through websites we interact with, answering machines in banking services, or through the use of "smart" devices and apps. Advanced applications of ML include image generation, music generation, and audio generation. In this chapter, we will use ML to control music in a more free or effective way, than the direct control of parameters through some kind of hardware or software controller.

7.1.2   How NNs work

Before we move on, we should clarify that NNs do not really learn anything. They comprise of placeholders for values and multiplication and addition coefficients that are adjusted by optimisation algorithms until they provide the desired output based on their input. Figure 7.1 illustrates a very small NN. This is a densely connected NN, or dense NN, where each neuron in one layer connects to all neurons in the next layer. Each connection transfers a value from the previous layer to the next, which is multiplied by a weight. The sum of all weighted values arriving at a neuron from a previous layer is added to a value called the bias, and is then fed to an activation function before it is transferred to the next layer. This sort of NN is called a feedforward NN. Figure 7.2 illustrates the inputs of one layer to the first neuron in the next layer, and that neuron's output. i is the input, w is the weight, and f is the activation function.

In supervised learning, which is what we will be dealing with in this chapter, while training a network, we provide sets of inputs with their associated outputs. These are expressed as numeric data. The network passes the input values through the process of multiplying them by the

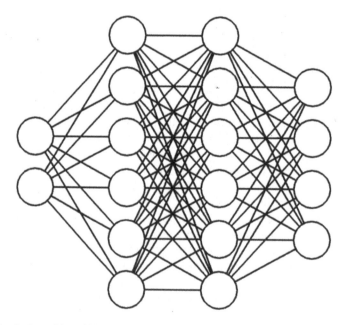

Figure 7.1 A simple dense Neural Network with two inputs, two hidden layers with six neurons each, and four outputs.

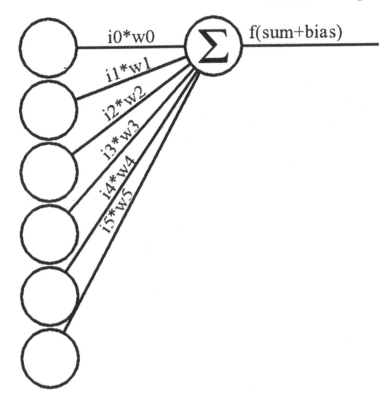

Figure 7.2 A forward pass from one layer to the first neuron in the next layer, and the output value of this neuron.

weights, adding them to the biases, and passing them through their activation function, through the entire network, at the end of which we get the result. This result is compared to the associated output of the input that has just been processed, and their difference, supplied by what is called the loss function, is stored. This value is called the loss.

Once this process is done, the network does a reverse process called back propagation, passing the values from the end to the beginning of the network. After the back propagation, the results are fed to an optimisation function, together with the loss, and through this function the weights and biases are adjusted slightly, so that in the next iteration, the network will provide an output that is a little bit less wrong. This process is repeated thousands of times, or even millions or more, until the output for any input in the training data is as close to the associated output as possible.

If the training process is successful, when the network is confronted with unseen input, it will provide correct output. We see that the terms "learn", "train", "predict", and other such terms used when talking about NNs, serve solely to help us communicate the ideas behind NNs. An NN does not learn, but only provides a correct set of coefficients so that numeric data that enter it, are properly processed and their result is what we expect it to be.

7.2   Available Python Modules for ML

Python is one of the most popular languages when it comes to ML. There are many modules for various ML tasks, including Scikit-learn, Theano, TensorFlow, Keras, and PyTorch. Throughout this chapter, we will use Keras. This module builds on top of TensorFlow, providing a simple API with minimal number of steps required to create NNs. It is intuitive and versatile, and it is used in all sorts of levels, from beginners to high professional and scientific level.

This module comes packaged with the TensorFlow module, as `tensorflow.keras`. To use Keras, you will have to install TensorFlow. Go ahead and install it by first typing `pip install --upgrade pip`, and then `pip install tensorflow`. We will also need numpy, a very efficient module for array manipulation and much more. We will need it to provide input and output data to our networks. To install it, type `pip install numpy`.

7.3   A Regression Neural Network

Two main uses of NNs are classification and regression. In classification, the NN is classifying its input. Classification use cases include classifying hand-written digits – the famous MNIST dataset – classifying images of clothes, fruit, animals, etc. When in regression mode, an NN outputs arbitrary values. A regression use case is housing prices. An NN is trained with house and area features, like number of rooms, crime rate in the area, accessibility to highways, and others, and it is then provided with the same features for a house that does not belong to its training dataset. The NN will predict the price for this house.

The housing prices use case uses an NN with many inputs and one output. We can reverse this and create an NN with a few inputs and many outputs. The inputs can be the coordinates of the computer's mouse pointer, and the outputs can be several parameters of a synthesizer. This way, we can control many parameters at once, with minimum input from us. This is the project we will build in this section.

Besides the Keras module, we will also need a module to monitor the mouse and keyboard of our computer. This module is pynput. Install it with `pip install pynput`. The code is longer than any other script in this book up to now, so we will first break it in smaller chunks. The full code will be provided at the end of this section.

Ingredients:

- Python3
- The Pyo module
- The wxPython module, if you want to use Pyo's GUI
- The keras module
- The numpy module
- The pynput module
- A text editor (preferably an IDE with Python support)
- A terminal window in case your editor does not launch Python scripts

Process:

We will start by importing all the necessary modules and defining some global variables, in a similar way we define macros in C++ with `#define`. This is done in Script 7.1. Macro definitions don't really exist in Python, but conventionally, global variables written in upper case are used this way. Lines 4 to 6 import modules and submodules for the neural network. Line 7 imports numpy with a shorter alias name, and line 8 imports the time and os modules. We will need the time module to create a polling mechanism for the mouse listener, and the os module to save our network model. The AUGSIZE macro defines the number of repetitions we will use to augment our training dataset. More on data augmentation later on. The NUMGENS macro defines the number of audio objects we will use. The POLLFR macro defines the polling frequency for the mouse listener. It is expressed in seconds.

Script 7.1 Importing the necessary modules and defining macros.

```
1   from pyo import *
2   import random
3   from pynput import mouse, keyboard
4   from tensorflow import keras
5   from tensorflow.keras.models import Sequential
6   from tensorflow.keras.layers import Dense
7   import numpy as np
8   import time, os
9
10  AUGSIZE = 200
11  NUMGENS = 4
12  POLLFR = 0.05
13
```

The next step is to create our neural network. This is done in Script 7.2. In this script we create a neural network with two inputs, two hidden layers with 64 neurons each, and 20 outputs. We create a Sequential object, which is what our MLP is, and start adding layers. We first add an Input () object from the keras module. Note that we could have imported this class the same way we imported Sequential (). If we did it this way though, we would get a name clash with Pyo's Input () class. Since we import Pyo with an asterisk, we import the keras module without an asterisk, and access this class with keras.Input (). The shape of our input will be a one-dimensional input with two neurons. The argument to the shape attribute must be a tuple, so we define the shape the way it is done in line 15.

Script 7.2 A Keras neural network with two inputs, two hidden layers with 64 neurons each, and 20 outputs.

```
14  nn = Sequential()
15  nn.add(keras.Input(shape=(2,)))
16  nn.add(Dense(64, activation="sigmoid"))
17  nn.add(Dense(64, activation="sigmoid"))
18  nn.add(Dense(NUMGENS*5, activation="linear"))
19
20  nn.compile(optimizer='adam',loss="mean_squared_error",
21              metrics=['accuracy'])
```

Once we create our input layer, we add the hidden layers. Both hidden layers have 64 neurons, and they both use the Sigmoid activation function. This function ensures that its output will always be between 0 and 1, no matter how large or small its input is. The output of this function reaches 0 when its input reaches negative infinity, and 1 when its input reaches positive infinity. When its input is 0, its output is 0.5. The last layer, the output layer, has 20 neurons and uses a linear function, which means that the output is identical to the input. Finally, we compile our model using the Adaptive Momentum (Adam) optimizer, a Mean Squared Error (MSE) loss function, and the accuracy from its metrics.

In this project, we will use one of the examples that come with the E-Pyo text editor to create sound, and modify it to fit our needs. The training dataset for our network will be the parameters of this example. The code of this example defines a class with three stacked sine wave oscillators, where each oscillator modulates the frequency of the next. In total we will control 20 parameters for this synthesizer. Our dataset will consist of mouse pointer coordinates and synthesizer parameters.

Before we start creating our dataset, we will define the synthesizer class. We will learn more about classes in the next chapter, but for now the minimal class definition in Script 7.3 should not look very unfamiliar. It is better to focus on the Pyo objects of the class in the __init__() method, and understand the FM synthesis, rather than trying to understand the Python class intricacies. In the next chapter, all these intricacies will make sense. The denoms list used in the setnew() method will be defined and explained later.

Script 7.3 The FM3 class.

```
22
23   s = Server().boot()
24
25   t = HarmTable([1, 0.1])
26
27   class FM3:
28       def __init__(self,fcar,ratio1,ratio2,ndx1,ndx2,out=0):
29           self.fcar = SigTo(fcar, time=POLLFR)
30           self.fmod = SigTo(ratio1, mul=fcar, time=POLLFR)
31           self.fmodmod = SigTo(ratio2,mul=self.fmod,time=POLLFR)
32           self.amod = SigTo(ndx1, mul=self.fmod, time=POLLFR)
33           self.amodmod=SigTo(ndx2,mul=self.fmodmod,time=POLLFR)
34           self.modmod = Sine(self.fmodmod, mul=self.amodmod)
35           self.mod = Sine(self.fmod+self.modmod, mul=self.amod)
36           self.car = Osc(t, fcar + self.mod, mul=0.2)
37           self.eq = EQ(self.car, freq=fcar, q=0.707, boost=-12)
38           self.out = DCBlock(self.eq).out(out)
39
40       def setnew(self, fcar, ratio1, ratio2, ndx1, ndx2):
41           self.fcar.setValue(float(fcar*denoms[0]))
42           self.fmod.setValue(float(ratio1))
43           self.fmod.setMul(float(fcar)*denoms[0])
44           self.fmodmod.setValue(float(ratio2)*denoms[2])
45           self.amod.setValue(int(ndx1*denoms[3]))
46           self.amodmod.setValue(int(ndx2*denoms[4]))
47
48       def setrand(self):
49           fcar = random.randrange(100, 300)
50           ratio1 = random.random()
51           ratio2 = random.random() * 3 + 1
52           ndx1 = random.randrange(1, 12)
53           ndx2 = random.randrange(1, 8)
54           self.fcar.setValue(fcar)
55           self.fmod.setValue(ratio1)
56           self.fmod.setMul(fcar)
57           self.fmodmod.setValue(ratio2)
```

```
58              self.amod.setValue(ndx1)
59              self.amodmod.setValue(ndx2)
60              return [fcar, ratio1, ratio2, ndx1, ndx2]
61
```

Once we define our class, we can create an object of it. We will create four objects, each with different parameters, following the example from Pyo Examples → Synthesis → 02_FM3.py. To use list comprehension, we will first create a list of lists, with the creation arguments for each object, and then we will create our object list. This is done in Script 7.4. Each sublist contains six values, but the last one concerns the output channel, defined as the out kwarg in the FM3 class. The first five arguments are the carrier frequency, the frequency ratio for the first modulator, the frequency ratio for the second modulator, and the two indexes, one for each modulator, following the FM() class paradigm. Since we have four FM3() objects in our list, we will have 20 parameters to control in total.

Script 7.4 Creating the FM3 object list.
```
62  param_lists = [[125.00, 0.33001, 2.9993, 8, 4, 0],
63                 [125.08, 0.33003, 2.9992, 8, 4, 1],
64                 [249.89, 0.33004, 2.9995, 8, 4, 0],
65                 [249.91, 0.33006, 2.9991, 8, 4, 1]]
66  fm3 = [FM3(*param_lists[i]) for i in range(NUMGENS)]
67
```

The next step is to create a few more global variables. These are shown in Script 7.5. poll_ stamp will be used to control the polling frequency of the mouse. As soon as the script is launched, we take a time stamp and store it. Then we define two variables for the mouse coordinates, and then lists to store and process the training data. The two None variables will be assigned to numpy arrays when we augment our data. The predicting Boolean is used to determine whether we will ask for a prediction when we move the mouse, and the denoms list items will be used as denominators to normalize our data to a range between 0 and 1. Neural networks work better with values in this range, so we will divide each output datum by the maximum value our output data can take.

Script 7.5 Global variables.
```
68  poll_stamp = time.time()
69
70  mouse_x = 0
71  mouse_y = 0
72
73  mouse_coords = []
74  synth_params = []
75  temp_synth_params = []
76  training_data= []
77  # create two variable to store numpy arrays later
78  training_input_data = None
79  training_output_data = None
80
81  # boolean to determine if we want to predict
82  predicting = False
```

```
83
84   # denominators for parameter normalization
85   denoms = [300, 1, 4, 12, 8]
86
```

Let's move on and create the objects and functions that will listen to the mouse and keyboard activity. This is done in Script 7.6. At the bottom of this script, we create the two listeners and assign a function for the mouse move input and a function for a key press input. These two functions are defined in this script. When we move the mouse, the on_move() function will be called with two arguments, the X and Y coordinates of the mouse pointer. In there, we define three global variables and we store the coordinates to the respective globals.

Script 7.6 The mouse and keyboard listeners and their functions.

```
129  def on_move(x, y):
130      global mouse_x, mouse_y, poll_stamp
131      mouse_x = x
132      mouse_y = y
133      if predicting:
134          # get current time stamp
135          new_stamp = time.time()
136          # make sure enough time has elapsed
137          if (new_stamp - poll_stamp) > POLLFR:
138              # update time stamp
139              poll_stamp = new_stamp
140              # create a numpy array with mouse coordinates
141              pred_input = np.full((1, 2), [x/1919, y/1079])
142              # ask network for predictions
143              predictions = nn(pred_input).numpy()
144              for i in range(NUMGENS):
145                  # call each FM object
146                  # with the correct array portion
147                  fm3[i].setnew(*predictions[0][i*5:(i*5)+5])
148
149  def on_press(key):
150      global temp_synth_params, predicting
151      try:
152          if key.char == 'r':
153              temp_synth_params = setrand()
154          elif key.char == 's':
155              synth_params.append(temp_synth_params)
156              mouse_coords.append([mouse_x, mouse_y])
157          elif key.char == 'd':
158              # when we press the d key
159              # we augment the training data
160              augment_data()
161              # we stop listening to the keyboard
162              keyboard_listener.stop()
163              # and we train the neural network
164              # using 10% for validation
165              nn.fit(training_input_data,
166                     training_output_data,
```

```
167                        batch_size=32, epochs=300,
168                        validation_split=.1)
169               print("training done!")
170               predicting = True
171      # except errors from certain keys
172      except AttributeError:
173           pass
174
175  mouse_listener = mouse.Listener(on_move=on_move)
176  mouse_listener.start()
177
178  keyboard_listener = keyboard.Listener(on_press=on_press)
179  keyboard_listener.start()
180
```

When we will be done with creating and augmenting our dataset, and training our network, we will be ready to ask our network for predictions. This is where the `predicting` variable will come in handy. Moving the mouse can cause this function to be called at a high rate, but asking for predictions is a demanding task. This is why we need to create a polling mechanism where we can control its frequency. For this to work, we have to take a time stamp whenever the `on_move()` function is called, and then compare its difference to the `poll_stamp` global variable. If this difference is greater than the `POLLFR` macro, then we can ask for a prediction. We will first update the `poll_stamp` variable and then get a prediction. To get a prediction we have to provide input data, which is the mouse pointer coordinates. The `Sequential()` class takes a numpy array as an input, among other data types. We create an array in line 141, with one row and two columns, and we pass the X and Y coordinates to its two values, normalised according to the resolution of the computer's screen.

The `Sequential()` class has a method called `predict()`, but that is too CPU expensive, and our script will lag. To avoid this, we simply pass our numpy array straight to the object, and call its `numpy()` method, to return a numpy array itself, otherwise it would return a `tensorflow.Tensor` datatype, which is not useful in our case. Then we iterate over our audio generators list and pass the respective portion of the array as an unpacked list. This happens in line 147. The numpy array returned by the neural network will have one row and 20 columns, essentially an array of 20 values included in another array. To address the first item of this array, we must type `predictions[0][0]`. In this line, we use the i variable to determine the start and end of the chunk of values we want. Using the asterisk, we unpack the array items we strip from the array, so the function we call will receive five arguments, instead of one list with five items.

If you go back to Script 7.3 and read the `setnew()` function, you will see that we multiply each of the input arguments with the respective item of the `denoms` list, to get the predicted values in the range we need. Since the neural network will be trained with values in the range between 0 and 1, its predictions will also be in this range. We remedy this by multiplying by the denominators used to normalise the training data.

The next function we need to discuss is `on_press()`. This function is called whenever we press a key on the computer's keyboard. To easily control this program, we will use the keys "r" for random, "s" for save, and "d" for done. When we press "r", a function that sets random values to the synthesizer will be called. This function, called `setrand()`, is defined in Script 7.7. It calls the method of the `FM3()` class with the same name, that sets random values to the five attributes of each object, and returns a list with these values.

Each of these lists is inserted in a parent list and returned. The returned list is stored in the `temp_synth_params` list.

Script 7.7 Function to set random values to the synthesizer.

```
87   def setrand():
88       l = []
89       for i in range(NUMGENS):
90           l.append(fm3[i].setrand())
91       return l
92
```

When we press the "s" key, in Script 7.6, the `temp_synth_params` list that has stored the last random value is appended to the `synth_params` list, and the mouse pointer coordinates are appended to the `mouse_coords` list. This way we create sets of training data, where the `mouse_coords` list will include the input data, and the `synth_params` list will include the output data.

When we are done, we can press the "d" key. Our data will be augmented with the function in Script 7.8, we will stop listening to the computer's keyboard, and we will train our network by calling its `fit()` method. This method takes the input and output data as arguments, along with some other kwargs. In NNs, an epoch is one iteration of all the training data, so the `epochs` kwargs sets the number of iterations for the training. NNs are often trained in batches, where the training data are split in groups to fit the number of batches we set via the `batch_size` kwarg. Finally, the `validation_split` kwarg sets whether we will use part of the training data to validate the network, and how much of this data will be used. Once the training is done, we print it on the console, and we set the `predicting` Boolean to true, so we can start asking for predictions when we move the mouse. Note that all the above happen within a `try/except` block. This is because some keys, like the space bar, raise an `AttributeError`, so we want to escape that, in case we accidentally hit this key.

Script 7.7 has already been explained, but we need to point out the line numbering, which is smaller than that in Script 7.6. This is because the `setrand()` function is called by Script 7.6, so it must be defined before that. If you type the code along with reading this explanation, make sure that you write Scripts 7.7 and 7.8 before Script 7.6.

Let's now move on to the last bit, Script 7.8. This is where we augment our training data. This script is written so that we assign one sound for each corner of the computer's screen, and perhaps one more for the centre. Five sets of training data are very little for an NN. Training datasets for NNs are orders of magnitude greater than this, so we need to somehow expand our dataset. This is where data augmentation comes in. What the `augment_data()` function in Script 7.8 does, apart from normalising both input and output data, is to create small variations on the input data to create a larger dataset from the initial one with five entries. The variations are random numbers from 0 to 19, for every X and Y in our dataset.

Script 7.8 Data augmentation function.

```
93   def augment_data():
94       global training_input_data, training_output_data
95       training_input_data = np.empty((len(mouse_coords)*\
96                               AUGSIZE,2))
97       training_output_data=np.empty((len(mouse_coords)*\
98                               AUGSIZE, NUMGENS*5))
99       for i in range(len(mouse_coords)):
```

```
100        for j in range(AUGSIZE):
101            training_data.append([])
102            mouse_x_rand = mouse_coords[i][0] + \
103                            random.randrange(20)
104            mouse_y_rand = mouse_coords[i][1] + \
105                            random.randrange(20)
106            # normalize according to screen resolution
107            mouse_x_rand /= 1919
108            mouse_y_rand /= 1079
109            # make sure values don't exceed 1
110            if mouse_x_rand > 1.0: mouse_x_rand = 1.0
111            if mouse_y_rand > 1.0: mouse_y_rand = 1.0
112            training_data[(i*AUGSIZE)+j].append(mouse_x_rand)
113            training_data[(i*AUGSIZE)+j].append(mouse_y_rand)
114            # normalize target output based on denoms list
115            for group in synth_params[i]:
116                for n, param in enumerate(group):
117                    training_data[(i*AUGSIZE)+\
118                                    j].append(param/denoms[n])
119        # shuffle the training data set for better training
120        random.shuffle(training_data)
121        # then separate the input from the output data
122        for ndx1, item in enumerate(training_data):
123            for ndx2, num in enumerate(item):
124                if ndx2 < 2:
125                    training_input_data[ndx1][ndx2] = num
126                else:
127                    training_output_data[ndx1][ndx2-2] = num
128
```

It is important to note that we must not create variations for the output data, and all variations of the original input data for every entry should be associated with the same output data of that entry. This is why in the nested loop in line 115, we don't augment our output data, but we only normalise it. To do that, we divide each of its items by the corresponding item of the denoms list. If you compare the items of the demons list in Script 7.5, with the ranges of the random values in the setrand() method of the FM3() class in Script 7.3, you will see that they are the same. Thus, by dividing by the items of the denoms list, we normalise our target outputs to a range between 0 and 1. Make sure to change the values in lines 107 and 108, in case your computer's screen has a different resolution.

When augmenting our data is over, we end up with a list of 1000 lists, 200 items per sample (in case we store a sample for the centre of the screen too). Each of these 1,000 lists contains 22 items, the two mouse coordinates, and 20 target outputs. The Sequential() class takes the input and output data separately, so we need to separate them. We could have stored them separately in the augmentation process, but before we train our network, we must shuffle our dataset.

Without shuffling, the first 200 entries will concern one corner of the screen, the second 200 entries another corner, and so on. With such a dataset, the network will "learn" very well how to predict values for one corner, but as soon as the training dataset enters another corner, the network will "forget" the state of the previous corner. To avoid this, we shuffle our dataset, so that our network "learns" all entries simultaneously during training. To shuffle easily, we keep both input and output data in one list, and we shuffle the parent list with the 1,000 lists. This is done in line 120.

After we shuffle our dataset, we separate the input from the output data and store them separately in numpy arrays. These arrays are the `training_input_data` and `training_output_data` variables that were initially defined as None. We define them as two-dimensional numpy arrays, with 1,000 columns (derived by multiplying the size of the `mouse_coords` list by the `AUGSIZE` macro), two columns for the input, and 20 columns for the output. We then run a nested loop in line 122 and split the values from the `training_data` and store them to these two arrays.

One last thing we need to do before we conclude our code is a way to save our network, once it has been properly trained. This is shown in Script 7.9. The `save_model()` function takes a string with the name of the network. By calling `os.getcwd()`, we get the current working directory – the directory where this script is saved – and add to it the name we want to give to our network. Then we print a message saying that the network has been saved successful. Finally we call the `gui()` method of the Pyo server to start our script and keep it alive.

Script 7.9 Saving the network and staring the Pyo server.
```
181   # when done training we can save the model
182   def save_model(net_name):
183       path_to_file = os.getcwd() + "/" + net_name
184       nn.save(path_to_file)
185       print(f"saved network model in {path_to_file}")
186
187   s.gui(locals())
```

The full code is shown in Script 7.10. When you run it, you might get some warning messages, depending on how you installed Tensorflow, but these are harmless. Hit the "r" key on your keyboard until you get a sound that you like, and then move your mouse pointer to one corner of the screen and press the "s" key. Repeat this process for all four corners and the centre of the screen. Then hit the "d" key to train the network. As the training progresses, you will see lines like the following printed on your monitor.

```
Epoch 255/300
29/29 [==============================] - 0s 4ms/step - loss: 1.3424e-
05 - accuracy: 0.9300 - val_loss: 1.3914e-05 - val_accuracy: 0.9400
```

We can see the epoch count in the first line and the sample count of each batch in the second (29 from 29), followed by some information about the time it took for each step of this process, and then we get the loss and accuracy of the training and the same values of the validation process. In the line above, the loss is a very low value expressed in what is called scientific notation. In this notation, the value 1.3914e-05 is 0.000013914, which results from multiplying 1.3914 by 10 raised to the negative 5th power. The accuracy value in this line is quite high, as 1.0 is the highest it can get. If the accuracy value is not high enough, or the loss value is quite high, the network will not have trained well. This might be due to the training dataset. In this case, exit and re-launch the script and repeat the training process.

Don't be fooled if your accuracy reaches 1.0. This doesn't mean that the network has reached a perfect state, but rather that it has "memorised" the training data, but hasn't generalised. This means that when it is confronted with unseen data, its predictions will probably not be very precise. Overall, we want an NN to have a high accuracy value, but we also want it to generalise, so its internal state is applicable to any input we give it.

Script 7.10 The full code of the regression neural network training.

```
1   from pyo import *
2   import random
3   from pynput import mouse, keyboard
4   from tensorflow import keras
5   from tensorflow.keras.models import Sequential
6   from tensorflow.keras.layers import Dense
7   import numpy as np
8   import time, os
9
10  AUGSIZE = 200
11  NUMGENS = 4
12  POLLFR = 0.05
13
14  nn = Sequential()
15  nn.add(keras.Input(shape=(2,)))
16  nn.add(Dense(64, activation="sigmoid"))
17  nn.add(Dense(64, activation="sigmoid"))
18  nn.add(Dense(NUMGENS*5, activation="linear"))
19
20  nn.compile(optimizer='adam',loss="mean_squared_error",
21              metrics=['accuracy'])
22
23  s = Server().boot()
24
25  t = HarmTable([1, 0.1])
26
27  class FM3:
28      def __init__(self,fcar,ratio1,ratio2,ndx1,ndx2,out=0):
29          self.fcar = SigTo(fcar, time=POLLFR)
30          self.fmod = SigTo(ratio1, mul=fcar, time=POLLFR)
31          self.fmodmod=SigTo(ratio2,mul=self.fmod,time=POLLFR)
32          self.amod = SigTo(ndx1, mul=self.fmod, time=POLLFR)
33          self.amodmod=SigTo(ndx2,mul=self.fmodmod,time=POLLFR)
34          self.modmod = Sine(self.fmodmod, mul=self.amodmod)
35          self.mod = Sine(self.fmod+self.modmod, mul=self.amod)
36          self.car = Osc(t, fcar + self.mod, mul=0.2)
37          self.eq = EQ(self.car, freq=fcar, q=0.707, boost=-12)
38          self.out = DCBlock(self.eq).out(out)
39
40      def setnew(self, fcar, ratio1, ratio2, ndx1, ndx2):
41          self.fcar.setValue(float(fcar*denoms[0]))
42          self.fmod.setValue(float(ratio1))
43          self.fmod.setMul(float(fcar)*denoms[0])
44          self.fmodmod.setValue(float(ratio2)*denoms[2])
45          self.amod.setValue(int(ndx1*denoms[3]))
46          self.amodmod.setValue(int(ndx2*denoms[4]))
47
```

```
48      def setrand(self):
49          fcar = random.randrange(100, 300)
50          ratio1 = random.random()
51          ratio2 = random.random() * 3 + 1
52          ndx1 = random.randrange(1, 12)
53          ndx2 = random.randrange(1, 8)
54          self.fcar.setValue(fcar)
55          self.fmod.setValue(ratio1)
56          self.fmod.setMul(fcar)
57          self.fmodmod.setValue(ratio2)
58          self.amod.setValue(ndx1)
59          self.amodmod.setValue(ndx2)
60          return [fcar, ratio1, ratio2, ndx1, ndx2]
61
62   param_lists = [[125.00, 0.33001, 2.9993, 8, 4, 0],
63                  [125.08, 0.33003, 2.9992, 8, 4, 1],
64                  [249.89, 0.33004, 2.9995, 8, 4, 0],
65                  [249.91, 0.33006, 2.9991, 8, 4, 1]]
66   fm3 = [FM3(*param_lists[i]) for i in range(NUMGENS)]
67
68   poll_stamp = time.time()
69
70   mouse_x = 0
71   mouse_y = 0
72
73   mouse_coords = []
74   synth_params = []
75   temp_synth_params = []
76   training_data= []
77   # create two variable to store numpy arrays later
78   training_input_data = None
79   training_output_data = None
80
81   # boolean to determine if we want to predict
82   predicting = False
83
84   # denominators for parameter normalization
85   denoms = [300, 1, 4, 12, 8]
86
87   def setrand():
88       l = []
89       for i in range(NUMGENS):
90           l.append(fm3[i].setrand())
91       return l
92
93   def augment_data():
94       global training_input_data, training_output_data
95       training_input_data = np.empty((len(mouse_coords)*\
96                                       AUGSIZE,2))
97       training_output_data=np.empty((len(mouse_coords)*\
98                                       AUGSIZE, NUMGENS*5))
99       for i in range(len(mouse_coords)):
```

```
100             for j in range(AUGSIZE):
101                 training_data.append([])
102                 mouse_x_rand = mouse_coords[i][0] + \
103                             random.randrange(20)
104                 mouse_y_rand = mouse_coords[i][1] + \
105                             random.randrange(20)
106                 # normalize according to screen resolution
107                 mouse_x_rand /= 1919
108                 mouse_y_rand /= 1079
109                 # make sure values don't exceed 1
110                 if mouse_x_rand > 1.0: mouse_x_rand = 1.0
111                 if mouse_y_rand > 1.0: mouse_y_rand = 1.0
112                 training_data[(i*AUGSIZE)+j].append(mouse_x_rand)
113                 training_data[(i*AUGSIZE)+j].append(mouse_y_rand)
114                 # normalize target output based on denoms list
115                 for group in synth_params[i]:
116                     for n, param in enumerate(group):
117                         training_data[(i*AUGSIZE)+\
118                                     j].append(param/denoms[n])
119         # shuffle the training data set for better training
120         random.shuffle(training_data)
121         # then separate the input from the output data
122         for ndx1, item in enumerate(training_data):
123             for ndx2, num in enumerate(item):
124                 if ndx2 < 2:
125                     training_input_data[ndx1][ndx2] = num
126                 else:
127                     training_output_data[ndx1][ndx2-2] = num
128
129 def on_move(x, y):
130     global mouse_x, mouse_y, poll_stamp
131     mouse_x = x
132     mouse_y = y
133     if predicting:
134         # get current time stamp
135         new_stamp = time.time()
136         # make sure enough time has elapsed
137         if (new_stamp - poll_stamp) > POLLFR:
138             # update time stamp
139             poll_stamp = new_stamp
140             # create a numpy array with mouse coordinates
141             pred_input = np.full((1, 2), [x/1919, y/1079])
142             # ask network for predictions
143             predictions = nn(pred_input).numpy()
144             for i in range(NUMGENS):
145                 # call each FM object
146                 # with the correct array portion
147                 fm3[i].setnew(*predictions[0][i*5:(i*5)+5])
148
149 def on_press(key):
150     global temp_synth_params, predicting
151     try:
```

```
152              if key.char == 'r':
153                  temp_synth_params = setrand()
154              elif key.char == 's':
155                  synth_params.append(temp_synth_params)
156                  mouse_coords.append([mouse_x, mouse_y])
157              elif key.char == 'd':
158                  # when we press the d key
159                  # we augment the training data
160                  augment_data()
161                  # we stop listening to the keyboard
162                  keyboard_listener.stop()
163                  # and we train the neural network
164                  # using 10% for validation
165                  nn.fit(training_input_data,
166                          training_output_data,
167                          batch_size=32, epochs=300,
168                          validation_split=.1)
169                  print("training done!")
170                  predicting = True
171          # except errors from certain keys
172          except AttributeError:
173              pass
174
175  mouse_listener = mouse.Listener(on_move=on_move)
176  mouse_listener.start()
177
178  keyboard_listener = keyboard.Listener(on_press=on_press)
179  keyboard_listener.start()
180
181  # when done training we can save the model
182  def save_model(net_name):
183      path_to_file = os.getcwd() + "/" + net_name
184      nn.save(path_to_file)
185      print(f"saved network model in {path_to_file}")
186
187  s.gui(locals())
```

Once you are done with training the network and playing with it, call the save_model()
function with the name you want the network to be saved as. This will create a directory with
this name in the same directory where your script is located. Script 7.11 shows the code for
loading a saved network. To use a saved network, we don't need to listen to the keyboard of the
computer, and we don't need to train the network any more, so we omit the keyboard submodule, and we only import Keras.

Script 7.11 Loading a saved neural network model.
```
1  from pyo import *
2  from pynput import mouse
3  from tensorflow import keras
4  import numpy as np
5  import time
6  import sys, os
```

```
7
8   path_to_model = None
9
10  if len(sys.argv) == 1:
11      path_to_model = input("Provide path to Keras model: ")
12      path_to_model = os.getcwd() + "/" + path_to_model
13  elif len(sys.argv) > 2:
14      print("This script takes one argument only")
15      exit()
16  else:
17      path_to_model = os.getcwd() + "/" + sys.argv[1]
18
19  NUMGENS = 4
20  POLLFR = 0.05
21
22  poll_stamp = time.time()
23
24  # denominators for parameter normalization
25  denoms = [300, 1, 4, 12, 8]
26
27  s = Server().boot()
28
29  t = HarmTable([1, 0.1])
30
31  param_lists = [[125.00, 0.33001, 2.9993, 8, 4, 0],
32                 [125.08, 0.33003, 2.9992, 8, 4, 1],
33                 [249.89, 0.33004, 2.9995, 8, 4, 0],
34                 [249.91, 0.33006, 2.9991, 8, 4, 1]]
35
36
37  class FM3:
38      def __init__(self,fcar,ratio1,ratio2,ndx1,ndx2,out=0):
39          self.fcar = SigTo(fcar, time=POLLFR)
40          self.fmod = SigTo(ratio1, mul=fcar, time=POLLFR)
41          self.fmodmod = SigTo(ratio2,mul=self.fmod,time=POLLFR)
42          self.amod = SigTo(ndx1, mul=self.fmod, time=POLLFR)
43          self.amodmod=SigTo(ndx2,mul=self.fmodmod,time=POLLFR)
44          self.modmod = Sine(self.fmodmod, mul=self.amodmod)
45          self.mod = Sine(self.fmod+self.modmod, mul=self.amod)
46          self.car = Osc(t, fcar + self.mod, mul=0.2)
47          self.eq = EQ(self.car, freq=fcar, q=0.707, boost=-12)
48          self.out = DCBlock(self.eq).out(out)
49
50      def setnew(self, fcar, ratio1, ratio2, index1, index2):
51          self.fcar.setValue(float(fcar*denoms[0]))
52          self.fmod.setValue(float(ratio1))
53          self.fmod.setMul(float(fcar)*denoms[0])
54          self.fmodmod.setValue(float(ratio2)*denoms[2])
55          self.amod.setValue(int(index1*denoms[3]))
56          self.amodmod.setValue(int(index2*denoms[4]))
57
58
```

```
59   # create the FM objects list
60   fm3 = [FM3(*param_lists[i]) for i in range(NUMGENS)]
61
62
63   def on_move(x, y):
64       global poll_stamp
65       # get current time stamp
66       new_stamp = time.time()
67
68       # make sure enough time has elapsed
69       if (new_stamp - poll_stamp) > POLLFR:
70           # update time stamp
71           poll_stamp = new_stamp
72           # create a numpy array with mouse coordinates
73           pred_input = np.full((1, 2), [x/1919, y/1079])
74           # ask network for predictions
75           predictions = nn(pred_input).numpy()
76           # call each FM object with the correct array portion
77           for i in range(NUMGENS):
78               fm3[i].setnew(*predictions[0][i*5:(i*5)+5])
79
80   # load the saved model
81   nn = keras.models.load_model(path_to_model)
82   print(f"loaded {path_to_model}")
83
84   mouse_listener = mouse.Listener(on_move=on_move)
85   mouse_listener.start()
86
87   s.gui(locals())
```

Most of the code is identical to the code in Script 7.10. We do include some new elements though. We have imported the sys module, so we can provide a command line argument. In line 10 we check the number of arguments by checking the length of the argv list (arguments vector) of the sys module. The first argument passed to the script is the name of the script, so the length of this list will be at least 1, even if we pass no arguments when we run it. If the length is 1, we know we haven't provided any arguments, and we prompt the user to provide the name of a saved model to load. The input() function is a built-in function that gives access to the shell running this Python script. Note that it has a lowercase initial i, in contrast to Pyo's Input() class that takes input from a microphone. If we provide more than one argument, then the length of the argv list will be greater than 2, and we inform the user that this script takes only one argument, and we exit. If we provide one argument, we store it with the directory of the running script prepended to it.

The other difference between these two scripts is that we have removed the setrand() method in the FM3() class, since we don't need to set any random values. In the on_move() function, we omit the predicting Boolean we used earlier, as we want to ask for predictions as soon as we launch the script. In line 81 we load the saved model and we print a message to the console stating that the model has been loaded successfully. To run this script, navigate to its directory in a terminal window and type the following:

```
python script_name.py model_name
```

Replace `script_name` with the name you saved the script with, and `model_name` with the name you saved the NN model with. As soon as the script launches, you should be able to transform the sound by moving the mouse pointer.

Documentation Pages:

`EQ()`: https://belangeo.github.io/pyo/api/classes/filters.html#eq
`DCBlock()`: https://belangeo.github.io/pyo/api/classes/filters.html#dcblock

7.4   A Classification Neural Network

We will now see how we can use a classification NN in a musical context. In this project we will classify vowels we will pronounce ourselves. We will build an NN that will take input from a microphone and it will be able to determine which vowel we are pronouncing. To do this, we will need to extract the Mel Frequency Cepstrum Coefficient (MFCC) from the microphone input, and feed it to our NN. The MFCC – Cepstrum is an anagram of the word Spectrum, pronounced with a hard C –, is a feature vector often used in speech signal processing (Zhang *et al.*, 2010). The information this vector provides can help us distinguish certain characteristics in sound, like distinct sounds of the human voice, or different instruments in the same input signal. The Python module we will use to extract the MFCC is called librosa, mentioned in the first chapter of this book. Go on and install it by typing `pip install librosa`.

Ingredients:

- Python3
- The Pyo module
- The wxPython module, if you want to use Pyo's GUI
- The keras module
- The numpy module
- The librosa module
- A text editor (preferably an IDE with Python support)
- A terminal window in case your editor does not launch Python scripts

Process:

We will again break the code in smaller chunks, as there are a few new concepts concentrated, that we need to discuss. We will start with importing all the necessary modules and defining some macros and a list with the vowels we will be classifying. All this is done in Script 7.12.

Script 7.12 Importing the necessary modules and creating macros and a vowel list.

```
1   from pyo import *
2   import random
3   import librosa
4   from tensorflow.keras.models import Sequential
5   from tensorflow.keras.layers import Dense
6   import numpy as np
7   import os
```

```
 8
 9   NUM_MFCC = 13
10   PATTHRESH = 0.01
11   CONF = 0.8
12
13   vowels = ["a", "o", "i", "e", "u"]
14
```

You might have noticed that we didn't import the keras module as a whole, like we did in the previous project, so we could create a `keras.Input()` object. The three macro definitions will be discussed later on. The vowels list includes the five vowels we will classify with our NN.

In Script 7.13, we create our NN. We see another way we can create it, without the `keras.Input()` class. Instead of adding this class at the beginning of the NN, we can immediately add a Dense layer, and specify its shape with the `input_shape` kwarg. This is essentially the same as writing `nn.add(keras.Input(shape=(NUM_MFCC,))`. The number of neurons in the input layer is defined by the `NUM_MFCC` macro we defined in Script 7.12. We will use this macro again, when we will extract the MFCC.

Script 7.13 Creating the Neural Network.
```
15   nn = Sequential()
16   nn.add(Dense(32, input_shape=(NUM_MFCC,), activation="relu"))
17   nn.add(Dense(32, activation="relu"))
18   nn.add(Dense(5, activation="softmax"))
19
20   nn.compile(optimizer='adam',
21               loss='categorical_crossentropy',
22               metrics=['accuracy'])
23
```

In this NN, we use two new activation functions, the ReLU, and the Softmax activation functions. ReLU stands for Rectified Linear Units. This function clips its input if it is below zero, and passes it intact otherwise, so its output is always 0 or greater. It is simpler than the Sigmoid function, but it is also much cheaper CPU wise, and there are many occasions where this function works just fine, so we prefer this over to the Sigmoid function.

The Softmax activation function is one of the functions used for classification. This function outputs a probability distribution. This distribution contains confidence values for each class, summing up to 1. In our case, we will be classifying five vowels, so this function will output a list of five values. The index of the highest value in each prediction will be the class the NN "believes" is correct. This means that this class will get the highest confidence, among the five classes. The loss function we use is also new. We use the Categorical Cross-Entropy loss function, which compares a "ground truth" probability and a predicted distribution (Kinsley and Kukieła, 2020, p. 112).

Once we create our NN, we boot and start the Pyo server and create some of the Pyo objects we will use for this project. This is done in Script 7.14. Note that we call the `start()` method of the Server, because we will not use Pyo's GUI in this project, although, we might need it later on. We don't use Pyo's GUI because we will need to interact with our script, and it is easier if we provide a prompt where the user can type minimal information, rather than having to call functions from the Interpreter entry of Pyo's Server window.

Script 7.14 Create Pyo objects and a comparison global variable.

```
24   s = Server().boot()
25   s.start()
26
27   mic = Input()
28   tab = DataTable(s.getBufferSize())
29   tabfill = TableFill(mic, tab)
30
31   prev_vowel = -1
32
```

In line 27 we create an `Input()` object to receive input from a microphone, then a table where we will store the microphone's input, and a `TableFill()` object that will write the input of the microphone to the table. The size of the table is Pyo's buffer size, the number of samples in one block. Pyo's default buffer size is 256, unless you use the Jack audio server, which has a 1024 default size. If while running this script you cannot get good results, consider raising the block size. 1024 seems to work fine, so if you are not using Jack, you will might need to raise Pyo's default block size with the `buffersize` kwarg of `Server()`. In line 31 we create a global variable that we will use once our NN has been trained, to avoid printing predictions to the console repeatedly.

The next thing we need to do is to define a function that will provide the MFCC. This is done in Script 7.15. The librosa MFCC class takes NumPy arrays in its input, so we will need to convert the microphone signal values to a NumPy array. This is done in line 34, where we store the contents of the table as such an array. Once this is done, we call the `mfcc()` function of librosa's feature submodule. The `y` kwarg is the signal input, which is the NumPy array with the microphone values. Then we define how many MFCCs we want. The default is 20, but for our purposes, 13 is enough, so we lower the default value to keep our script lighter. We must also provide the sampling rate of our signal, and we do this by calling the `getSamplingRate()` method of Pyo's `Server()` class.

Script 7.15 Function that returns the MFCC.

```
33   def get_mfcc():
34       mic_array = np.asarray(tab.getBuffer())
35       mfcc = librosa.feature.mfcc(y=mic_array,n_mfcc=NUM_MFCC,
36                                   sr=s.getSamplingRate())
37       return mfcc
38
```

The next two functions are the training and predicting functions. These are shown in Script 7.16. There is nothing new to discuss for the training function, but there are some new points in the predicting function. Here is where we will use the `prev_vowel` global variable. In line 47 we get the MFCC by calling the `get_mfcc()` function as the second argument of the `np.full()` NumPy function. The first argument to the latter is the shape of the NumPy array we want to create, and the second is the values we want this new array to be filled with. The MFCC output is a list of NUM_MFCC (13) lists, where each list has three values. For our purposes, the first value of each of these lists is enough, so instead of storing the entire MFCC output, we store only its first column by using the syntax in line 47.

Script 7.16 The training and predicting functions.

```
39  def train_network(input_data, output_data):
40      nn.fit(input_data, output_data,
41          batch_size=32, epochs=100,
42          validation_split=.1)
43
44  def predict():
45      global prev_vowel
46      # get only first column of MFCC 2D array inside an array
47      pred_input = np.full((1, NUM_MFCC), get_mfcc()[:, 0])
48      prediction = nn.predict(pred_input, verbose=0)
49      predicted_vowel = np.argmax(prediction)
50      if prediction[0][predicted_vowel] < CONF:
51          pass
52      else:
53          if predicted_vowel != prev_vowel:
54              print(f"prediction: {vowels[predicted_vowel]} ")
55              prev_vowel = predicted_vowel
56
```

In line 48 we call the `predict()` method of our NN with the NumPy array containing the input data. We set the `verbose` kwarg to 0, so the NN doesn't print anything on the console and it doesn't get mixed with what we want to print ourselves. As already mentioned, the output of this NN is a list with confidence values for each class. To retrieve the actual predicted class, we need to find the index of the greatest value. We do this with NumPy's `argmax()` function.

The NN output is a list of five values inside another list. For this reason, in line 50, we test the highest confidence value by indexing both dimensions of the `prediction` array. In this line we test if the highest confidence is above a confidence threshold that we have set as a macro. A classification NN will produce output regardless of the input, even if the input doesn't make any sense in the context of the training of the network. This means that even if we provide noise as input, our NN will still provide output, though its confidence values will be rather flat, around 1/5 each. These values will not be equal though, and one of them will still be the greatest, which will result in the predicted class. To avoid this, we set a threshold and we test the highest confidence value against it. If the predicted class confidence is lower than this threshold, we just `pass` and exit the function.

If the predicted class confidence is above the threshold, we test if the prediction index is different than the last prediction index, and if it is, we print the corresponding item of the `vowels` list, so we print the actual vowel and not its index. We then update the `prev_vowel` global variable, so the test in line 53 works every time.

After we define these two functions, we need to define a few more things, shown in Script 7.17. The saving function in line 69 is identical to its corresponding function from the previous project. In lines 62 to 67 we create some Pyo objects that enable the script to predict vowels without any keyboard input from the user. The `Pattern()` object will call the `predict()` function when it is activated. In line 63 we create a `Follower()` object that outputs the mean amplitude of its first argument, which is the microphone input.

Script 7.17 Predictions polling and model saving functions, and Pyo polling and triggering objects.

```
57  def poll_predictions():
58      global prev_vowel
```

```
59        prev_vowel = -1
60        pat.play()
61
62   pat = Pattern(predict, time=.1)
63   follower = Follower(mic)
64   thresh1 = Thresh(follower, PATTHRESH, dir=0).stop()
65   thresh2 = Thresh(follower, PATTHRESH, dir=1).stop()
66   tf1 = TrigFunc(thresh1, poll_predictions)
67   tf2 = TrigFunc(thresh2, pat.stop)
68
69   def save_model(net_name):
70        path_to_file = os.getcwd() + "/" + net_name
71        nn.save(path_to_file)
72        print(f"saved network model in {path_to_file}")
73
```

Lines 64 and 65 create two `Threshold()` objects. These take a PyoObject in their first argument and output a trigger whenever the signal of this input crosses a threshold value set as the second argument. This threshold is defined by the `PATTHRESH` macro. The `dir` attribute sets whether the trigger will occur when the input signal crosses the threshold value from below to above – that's when `dir` is set to 0 – of from above to below – when `dir` is set to 1. We immediately `stop()` both objects, because we must first create the training dataset and train our network. These objects will be activated once the training has finished.

Lines 66 and 67 create two `TrigFunc()` objects that trigger a function whenever the get a trigger signal in their first input. The first of these two objects calls the `poll_predictions()` function. This function resets the `prev_vowel` to -1 and activates the `Pattern()` object. We need to reset `prev_vowel`, so that we get a prediction if we pronounce the same vowel twice, with a pause in between. The second `TrigFunc()` object deactivates the `Pattern()` object. With this chunk of code we activate the predictions polling only when there is audio coming in the microphone, and disable it when there is silence, saving some CPU.

The last bit we need to discuss is shown in Script 7.18. Here we use the `if __name__ == "__main__"`: test we saw in Chapter 2. In there, we create a local variable to store the index of the vowel we will set manually each time, for the training dataset. Then we create a list that will store all the MFCCs together with the index of the vowel they represent. Then we create a string that will be used as a prompt, when the whole script is run. We prompt the user to either provide one of the vowels from the `vowels` list, or hit the return key to get an MFCC of a snapshot – essentially a sample block – 't' to train the NN when the dataset is ready, or 'q' to exit the script. Then we create a Boolean which is used in the `while` loop in line 80, and we enter this loop.

Script 7.18 The if __name__ == "__main__": code chunk.

```
74   if __name__ == "__main__":
75        vowel = None
76        mfccs = []
77        prompt = "Provide vowel or hit return for snapshot, "
78        prompt += "'t' to train, or 'q' to quit: "
79        training = True
80        while training:
81             user_input = input(prompt)
```

```
82              if user_input == "q":
83                  exit()
84              elif user_input in vowels:
85                  vowel = vowels.index(user_input)
86              elif user_input == "":
87                  if vowel is None:
88                      print("No vowel provided yet!")
89                  else:
90                      mfccs.append([get_mfcc(), vowel])
91              elif user_input == "t":
92                  if len(mfccs) == 0:
93                      print("No MFCCs stored yet!")
94                  else:
95                      in_data = np.empty((len(mfccs),
96                                          mfccs[0][0].shape[0]))
97                      out_data=np.empty((len(mfccs),len(vowels)))
98                      random.shuffle(mfccs)
99                      for i in range(len(mfccs)):
100                         # get only first column of MFCC
101                         in_data[i] = mfccs[i][0][:, 0]
102                         # create one-hot array
103                         for j in range(len(vowels)):
104                             out_data[i][j] = 0
105                         out_data[i][mfccs[i][1]] = 1
106                     train_network(in_data, out_data)
107                     training = False
108
109         thresh1.play()
110         thresh2.play()
111         prompt1 = "Training done, now detecting vowels . . . "
112         prompt2 = "Is this a name for saving the model? (y/n): "
113         while True:
114             user_input = input(prompt1)
115             if user_input == "q":
116                 exit()
117             else:
118                 answer = input(prompt2)
119                 if answer == "y":
120                     save_model(user_input)
```

In this loop, we print the prompt to the console and wait for a response from the user. The first thing we will need to do as the user of this script, is to provide the vowel we want to store MFCCs of. Once we provide a vowel and hit return, if it is found in the vowels list, its index from this list will be stored and the while loop will resume, printing the prompt again. The next thing we will need to do is to pronounce this vowel into the microphone and at the same time hit the return key. This will result in the test in line 86 to pass and execute its code. If we haven't provided a vowel yet, we will be notified accordingly, nothing else will happen, and the while loop will resume, otherwise the MFCC of the microphone input together with the vowel index will be appended to the mfccs list.

We have to repeat this process for all five vowels. Once we are done, we can type 't'so the NN can be trained. If we haven't stored any data yet, line 93 notifies us, otherwise, in lines 95 and 97, we create the input and output data NumPy arrays, then we shuffle the training dataset, and pass the MFCCs to the `in_data` array, and the corresponding indexes of the vowels to the `out_data` array. We get the first column of the MFCC with the same syntax we used in line 47 in Script 7.16. For this project, we don't need to normalise our data to a range between 0 and 1. For the output dataset, we have to create what is called a one-hot array. Instead of storing a single value – the index of the vowel – we create an array with as many items as the number of classes – in this case five, as many as the vowels – and for each sample in the training dataset we set all the values to 0, and the value at the index of the class of this sample to 1.

Remember that the `mfccs` list contains as many items as the MFCCs we stored manually, and each item has two items, the MFCC and the index of the vowel. This is why we need to access the items of this list with two indexes, i and 0 or 1. Note also that the `in_data` and `out_data` arrays are two-dimensional, with rows as many as the items of our dataset, and 13 columns for the input data – the number of MFCCs – and five columns for the output data – the number of the vowels.

After we separate the input data from the output as NumPy arrays, we train the network. When the training is done, we set the `training` Boolean to `False`, so the `while` loop can exit and we can move on. We then activate the two `Threshold()` objects so we can start polling predictions when we provide input to the microphone. We create a new prompt and print that to the console, to notify the user that the script is now detecting vowels. We can still provide text input though, so we can either exit the script or save the network. Exiting happens in lines 83 and 116, either before or after the training. Saving the network happens once the training is over, in line 120, if we provide any text input other than single vowels or 'q'. Since it is possible to provide random input by accidentally pressing keys on the keyboard, we ask the user if the provided input is really a name to store the NN model. If we type 'y' for "yes", we save the model the same way we did in the previous project. The full code is shown in Script 7.19.

Script 7.19 The vowel classification script.

```
1   from pyo import *
2   import random
3   import librosa
4   from tensorflow.keras.models import Sequential
5   from tensorflow.keras.layers import Dense
6   import numpy as np
7   import os
8
9   NUM_MFCC = 13
10  PATTHRESH = 0.01
11  CONF = 0.8
12
13  vowels = ["a", "o", "i", "e", "u"]
14
15  nn = Sequential()
16  nn.add(Dense(32, input_shape=(NUM_MFCC,), activation="relu"))
17  nn.add(Dense(32, activation="relu"))
18  nn.add(Dense(5, activation="softmax"))
19
20  nn.compile(optimizer='adam',
21            loss='categorical_crossentropy',
```

```
22                    metrics=['accuracy'])
23
24 s = Server().boot()
25 s.start()
26
27 mic = Input()
28 tab = DataTable(s.getBufferSize())
29 tabfill = TableFill(mic, tab)
30
31 prev_vowel = -1
32
33 def get_mfcc():
34     mic_array = np.asarray(tab.getBuffer())
35     mfcc = librosa.feature.mfcc(y=mic_arrayn_mfcc=NUM_MFCC,
36                                 sr=s.getSamplingRate())
37     return mfcc
38
39 def train_network(input_data, output_data):
40     nn.fit(input_data, output_data,
41           batch_size=32, epochs=100,
42           validation_split=.1)
43
44 def predict():
45     global prev_vowel
46     # get only first column of MFCC 2D array inside an array
47     pred_input = np.full((1, NUM_MFCC), get_mfcc()[:, 0])
48     prediction = nn.predict(pred_input, verbose=0)
49     predicted_vowel = np.argmax(prediction)
50     if prediction[0][predicted_vowel] < CONF:
51         pass
52     else:
53         if predicted_vowel != prev_vowel:
54             print(f"prediction: {vowels[predicted_vowel]} ")
55             prev_vowel = predicted_vowel
56
57 def poll_predictions():
58     global prev_vowel
59     prev_vowel = -1
60     pat.play()
61
62 pat = Pattern(predict, time=.1)
63 follower = Follower(mic)
64 thresh1 = Thresh(follower, PATTHRESH, dir=0).stop()
65 thresh2 = Thresh(follower, PATTHRESH, dir=1).stop()
66 tf1 = TrigFunc(thresh1, poll_predictions)
67 tf2 = TrigFunc(thresh2, pat.stop)
68
69 def save_model(net_name):
70     path_to_file = os.getcwd() + "/" + net_name
71     nn.save(path_to_file)
72     print(f"saved network model in {path_to_file}")
73
```

```
74  if __name__ == "__main__":
75      vowel = None
76      mfccs = []
77      prompt = "Provide vowel or hit return for snapshot, "
78      prompt += "'t' to train, or 'q' to quit: "
79      training = True
80      while training:
81          user_input = input(prompt)
82          if user_input == "q":
83              exit().
84          elif user_input in vowels:
85              vowel = vowels.index(user_input)
86          elif user_input == "":
87              if vowel is None:
88                  print("No vowel provided yet!")
89              else:
90                  mfccs.append([get_mfcc(), vowel])
91          elif user_input == "t":
92              if len(mfccs) == 0:
93                  print("No MFCCs stored yet!")
94              else:
95                  in_data = np.empty((len(mfccs),
96                                      mfccs[0][0].shape[0]))
97                  out_data=np.empty((len(mfccs),len(vowels)))
98                  random.shuffle(mfccs)
99                  for i in range(len(mfccs)):
100                     # get only first column of MFCC
101                     in_data[i] = mfccs[i][0][:, 0]
102                     # create one-hot array
103                     for j in range(len(vowels)):
104                         out_data[i][j] = 0
105                     out_data[i][mfccs[i][1]] = 1
106                 train_network(in_data, out_data)
107                 training = False
108
109     thresh1.play()
110     thresh2.play()
111     prompt1 = "Training done, now detecting vowels . . . "
112     prompt2 = "Is this a name for saving the model? (y/n): "
113     while True:
114         user_input = input(prompt1)
115         if user_input == "q":
116             exit()
117         else:
118             answer = input(prompt2)
119             if answer == "y":
120                 save_model(user_input)
```

To create a dataset, you will need to store around 20 MFCCs for each vowel, or more. While storing these MFCCs, make small variations to the vowel sound by opening and closing your mouth slightly, and keep on hitting the return key to take many snapshots. Be careful as vowel sounds can intersect. An "ah" sound with your mouth too closed can be very close to an "oh" sound, and the other way round. The same applies to all vowels. Try to keep the vowel sounds as distinct as possible.

One last step you will need to take before running Script 7.19, is to determine the loudness of your microphone. In line 10 we define the PATTHRESH macro to 0.01. This means that after the NN has been trained, whenever the mean amplitude of the microphone input goes above this value, the script will start detecting vowels. Depending on your setup, it is very likely that this value needs to be changed. Run Script 7.20 first and check the printed values in your console. Set the PATTHRESH macro in Script 7.19 half way between the values you get when you don't speak into the microphone and when you do speak into it.

Script 7.20 Determine the mean amplitude values of the microphone input.
```
1   from pyo import *
2
3   s = Server().boot()
4
5   mic = Input()
6   follower = Follower(mic)
7   p = Print(follower, interval=0.5)
8
9   s.gui(locals())
```

As with the previous project, this script should be accompanied by a script that loads the saved model, so you don't have to repeat the training process every time. This is left as an exercise, included in the exercises section of this chapter. Also, we only classified vowels and printed the predictions to the console. There is much more we can do with this script, in a musical context. This is also left as an exercise, also included in the exercises section.

Apart from vowel sounds, the MFCC lets us distinguish between other types of sounds, like different instruments. Suppose that you are part of an ensemble with a few instruments, and depending on the instrument that plays, something different needs to happen. If the hardware setup does not allow for individual inputs to your sound card, but only a mixed return signal from the PA is possible, it could suffice if you train this little program to distinguish between the different instrument sounds.

Documentation Pages:

DataTable(): https://belangeo.github.io/pyo/api/classes/tables.html#datatable
TableFill(): https://belangeo.github.io/pyo/api/classes/tableprocess.html#tablefill
Follower(): https://belangeo.github.io/pyo/api/classes/analysis.html#follower
Thresh(): https://belangeo.github.io/pyo/api/classes/triggers.html#thresh
Print(): https://belangeo.github.io/pyo/api/classes/utils.html#print

7.5   A Few Notes on Neural Networks and Training Datasets

Generally speaking, when we talk about NNs, we usually think of large datasets that cover all possible scenarios in the context of the NN we want to train. In many fields where NNs are used, this is indeed the case and a necessity. When it comes to music though, things are a little bit different. Rebecca Fiebrink, a pioneer in Human-Computer Interaction (HCI), while researching on HCI with ML, she found that small datasets are sufficient and even preferred over larger ones (Fiebrink, 2010, p. 203). Despite this fact, in the first project of this chapter, we augmented our data by 200 times, going from a dataset of five samples to one with 1,000 samples. Still, in a wider ML context, the latter dataset is considered extremely small.

Another issue we must face when training NNs is the hardware we use. Due to the heavy work load required to train an NN, many ML libraries utilise the Graphics Processing Unit (GPU) of computers, instead of the CPU. This happens because the GPU contains thousands of cores, in contrast to the CPU which contains only four or eight or maybe a little bit more. Not all hardware GPUs though are supported by all ML libraries. It is likely that when you run any of the projects we built in this chapter, you get a message printed onto your console stating that Tensorflow has not been built with GPU support. Training NNs with small datasets enables us to go through this process even on machines that do not support GPU training, and still get satisfactory results.

Even with a small dataset though, we must be careful to provide samples that represent as many scenarios as possible. This means that we must provide the biggest variety possible in our input data. Take the vowel classification script as an example, where we must pronounce each vowel in different ways, to cover many possibilities. Overall, we should be careful to not overfit the NN on one hand – getting an accuracy of 1.0 is usually a sign of overfitting – but get an accuracy that is high enough and a loss value that is as low as possible. It might be necessary to go through the training process of an NN more than once, until you get a good result.

7.6 Conclusion

NNs have numerous applications, even only within the domain of music. There are many libraries available for various ML tasks, many of which are open-source and free to use. We have focused on the Keras module for Python because it builds on top of TensorFlow, a very effective module for NNs, but provides a user-friendly interface, enabling fast prototyping. This renders Keras useful to beginners up to a high professional level research. You are encouraged to check other modules too and compare them to Keras, both from a performance and an intuition perspective.

We have seen two different types of NNs for different music applications. This is only the beginning of what can turn out to be a continuous journey in AI, as we have only scratched the surface in a field of ongoing research and development. There are many more types of NNs, depending on the task at hand. From controlling multiple parameters with a handful of inputs, to generating raw audio or even symbolic music, like scores, NNs can prove to be very useful tools, assisting the creative process of making music.

This chapter is not intended to be an exhaustive resource for AI and NNs. You are encouraged to read other books, papers, articles, or any resource you can find that can provide insight to NNs. Even if you don't intend to write your own code for NNs, having gone through this chapter will help you demystify certain concepts of AI, understand software that incorporates NNs and use it in a sensible and effective way.

7.7 Exercises

7.7.1 *Exercise 1*

Write the accompanying script to Script 7.19, that loads a saved NN, to distinguish vowels.
Tip: Remove any code that deals with creating a training dataset and that trains the network. Copy any necessary bits from Script 7.11 from the first project of this chapter,

including the lines that import the necessary modules for loading a model. Make sure to remove lines that import modules that are not necessary. In this version, you can use `s.gui(locals())` instead of calling `s.start()`.

7.7.2 Exercise 2

Develop the script from exercise 1 to do something musical with the predictions.

Tip: You can use a `Selector()` and control its `voice` attribute with the network's predictions. Make sure you make the necessary conversions to pass the correct data type to `Selector()`'s voices.

7.7.3 Exercise 3

Switch the mouse pointer coordinates in Script 7.11 from the first project with input from an Arduino.

Tip: A joystick is a good replacement for the mouse or mouse pad, as it provides X and Y coordinates and it is used with one hand. Make sure the input to the NN is in the range between 0 and 1.

Bibliography

Fiebrink, R. (2010) *Real-time Human Interaction with Supervised Learning Algorithms for Music Composition and Performance*. PhD Thesis. Princeton University.

Gardner, M.W. and Dorling, S.R. (1998) 'Artificial neural networks (the multilayer perceptron): A review of applications in the atmospheric sciences', *Atmospheric Environment*, 32(14), pp. 2627–2636. Available at: https://doi.org/10.1016/S1352-2310(97)00447-0.

Kinsley, H. and Kukieła, D. (2020) *Neural Networks from Scratch in Python*. Available at: https://nnfs.io.

Priddy, K.L. and Keller, P.E. (2005) *Artificial Neural Networks: An Introduction*. SPIE Publications.

Zhang, W.-Q. *et al.* (2010) 'Perturbation analysis of mel-frequency cepstrum coefficients', in *2010 International Conference on Audio, Language and Image Processing*, pp. 715–718. Available at: https://doi.org/10.1109/ICALIP.2010.5685063.

8 Writing Your Own Classes in Python

In this chapter we will learn how to write our own classes to cover possible scenarios that are not covered by the array of Pyo classes. This chapter will help decypher certain concepts of Object-Oriented Programming (OOP), and will clear any confusion on certain terminology we have used so far. We will see why OOP provides flexibility and we will understand when we need to write classes of our own.

What You Will Learn

After you have read this chapter you will:

- Understand key-concepts of OOP
- Know the difference between a class, an object, and a method
- Be able to write your own classes to cover specific needs

8.1 What are Python Classes and How to Write One

We have already run into the terms *class*, *object*, *method*, *function*, and *variable*. Especially the first two terms have been often used inter-changeably. This is because a class defines a set of variables and functions that are packaged in a unified code structure, and an object is an instance of a class. Let's see this with some hands-on examples.

Ingredients:

- Python3
- A text editor (preferably an IDE with Python support)
- A terminal window in case your editor does not launch Python scripts

Process:

Suppose we have a class called `Person()`. By convention, class names in Python start with an uppercase letter and all words in the name of the class are concatenated and separated with uppercase letters, like `SineLoop()` and `SuperSaw()`. The opening and closing round brackets are used here to specify that this is a class. This imaginary class can store some

DOI: 10.4324/9781003386964-8

values that define a person, like age, gender, height, and weight. This class is pretty pointless until we create an instance of it and store these values. To create an instance, we can type the following:

```
jane = Person(age=30, gender="f", height=1.65, weight=58)
```

We have now created an instance of the class Person() and have provided values to all the attributes of the class. This instance is called jane, and it is an object of the class Person(). We can create another instance of this class – another object – with a different name and different values, like we do in the line below:

```
john = Person(age=24, gender="m", height=1.76, weight=71)
```

The two Person() objects, jane and john, are independent of each other and the values of one have no effect on the values of the other. Let's write the code for this class. The most basic structure of this class is shown in Script 8.1.

Script 8.1 Basic structure of a class.
```
1    class Person:
2         def __init__(self,age=25,gender=None,height=1.7,weight=65):
3             self._age = age
4             self._gender = gender
5             self._height = height
6             self._weight = weight
7             if self._gender is None:
8                 print("Gender hasn't been provided!")
9
```

When we create an object of a class, its __init__() method is called. If we want anything to happen on object instantiation, we have to define this method. Notice that we use the word self in many places. Apart from being prepended to variable names, it is also the first argument in any method of a class. This word – mind that it is not a keyword, it could be anything else, but by convention the word self is used throughout the Python community – refers to the instance of the class, so when we want to access a variable of an object of a class, or call one of its methods, the variable or method of the specific instance will be accessed, and not that of another object.

To better understand how self works in methods, suppose that we define a method that takes no arguments and we omit to place this word in the definition. When we call this method, we will get a TypeError saying that the method takes 0 arguments but 1 was given. An example of this is shown in Script 8.2. If we call this method with jane.wave_hello(), we will get the TypeError mentioned above, even though the method definition and the way we call it seem to agree. The word self has to be provided as the first argument when we define class methods, even though it is not included in the method call.

Script 8.2 A class method without the "self" argument.
```
10    def wave_hello():
11         print("hello!")
```

Also note that even though Python does not support private variables, which can be manipulated only through methods of a class, and not by immediately calling them, an underscore in the beginning of a variable is a convention to define variables that are treated as private. For example, even though it is possible to set a new value for any of the variables in Script 8.1, like it is done in the line below, it is considered to be bad practice, and it is avoided by Python programmers.

```
jane._age = 32
```

Further on, when we write classes with PyoObjects, we will see how to avoid this. Let's now see how to create a method that is useful in the context of classes. Script 8.3 includes the initial class definition with two more methods. If you run this script, you will get the following output:

```
info of friend john:
age: 24
gender: m
height: 1.75
weight: 71

info of friend jack:
this is not a friend
```

Classes encapsulate code in a very convenient way that enables us to abstract our code to a high degree. To fully grasp how the word self functions within a class, read the get_friend_ info() method carefully. We access the _friends list of the object whose method we call – in this case, jane – but then we access the variables of the object john, by using the argument friend in the printed F-strings, instead of self. This is like calling john._age and all the other variables in a similar way, which, within the john instance, are accessed as self._age, but in this case, self refers to john and not jane.

Script 8.3 Person class definition with methods and object instantiation.

```
1   class Person:
2       def __init__(self,age=25,gender=None,
3                         height=1.7,weight=65):
4           self._age = age
5           self._gender = gender
6           self._height = height
7           self._weight = weight
8           self._friends = []
9           if self._gender is None:
10              print("Gender hasn't been provided")
11
12      def add_friend(self, friend):
13          self._friends.append(friend)
14
15      def get_friend_info(self, friend):
16          if friend in self._friends:
17              print(f"age: {friend._age}")
18              print(f"gender: {friend._gender}")
19              print(f"height: {friend._height}")
20              print(f"weight: {friend._weight}")
21          else:
22              print("this is not a friend")
```

```
23
24   jane = Person(age=30, gender="f", height=1.65, weight=58)
25   john = Person(age=24, gender="m", height=1.75, weight=71)
26   jack = Person(age=40, gender="m", height=1.8, weight=78)
27
28   jane.add_friend(john)
29   print("info of friend john:")
30   jane.get_friend_info(john)
31   print()
32   print("info of friend jack:")
33   jane.get_friend_info(jack)
```

8.1.1   Inheritance in Classes

Python supports the concept of inheritance in classes, where a class that we define can inherit variables and methods from another class, called the super class or the parent class. The class that inherits is called the subclass, or child class. This feature is convenient when we want to have a base class with certain features that will be shared among other classes. By using inheritance, we can save time by writing the code to be shared among classes only once. Script 8.4 expands on Script 8.3 by adding two more methods to our Person() class, and adds two more classes that inherit from Person(), Employee() and Student().

Script 8.4 The Person() parent class with the Employee() and Student() child classes.
```
1    class Person:
2        def __init__(self, age=25, gender=None,
3                     height=1.7, weight=65):
4            self._age = age
5            self._gender = gender
6            self._height = height
7            self._weight = weight
8            self._friends = []
9            self._relatives = []
10       if self._gender is None:
11           print("Gender hasn't been provided")
12
13       def add_friend(self, friend):
14           self._friends.append(friend)
15
16       def add_relative(self, relative):
17           self._relatives.append(relative)
18
19       def get_friend_info(self, friend):
20           if friend in self._friends:
21               print(f"age: {friend._age}")
22               print(f"gender: {friend._gender}")
23               print(f"height: {friend._height}")
24               print(f"weight: {friend._weight}")
25           else:
26               print("this is not a friend")
27
28       def get_relative_info(self, relative):
```

```
29              if relative in self._relatives:
30                  print(f"age: {relative._age}")
31                  print(f"gender: {relative._gender}")
32                  print(f"height: {relative._height}")
33                  print(f"weight: {relative._weight}")
34              else:
35                  print("this is not a relative")
36
37
38  class Employee(Person):
39      def __init__(self, age=25, gender=None, height=1.7,
40                      weight=65, firm=None):
41          super().__init__(age, gender, height, weight)
42          self._firm = firm
43          self._post = None
44
45      def add_post(self, post):
46          self._post = post
47
48      def get_post(self):
49          if self._post is not None:
50              print(f"this employee's post is {self._post}")
51          else:
52              print("this employee doesn't have a post yet")
53
54
55  class Student(Person):
56      def __init__(self, age=25, gender=None, height=1.7,
57                      weight=65, school=None):
58          super().__init__(age, gender, height, weight)
59          self._school = school
60          self._grades = {}
61
62      def set_grades(self, semester, grade):
63          self._grades[semester] = grade
64
65      def get_grades(self, semester):
66          if semester not in self._grades.keys():
67              print(f"grade for {semester} semester \
68                      has not been provided")
69          else:
70              print(f"{semester} grade: \
71                      {self._grades[semester]}")
72
73
74  jane = Student(age=15,gender="f",height=1.6,
75                  weight=50,school="Freedonia High")
76  john = Employee(age=34,gender="m",height=1.75,
77                  weight=71,firm="Freedo Realty")
78
79  jane.add_relative(john)
80  print("info of relative john:")
81  jane.get_relative_info(john)
```

To write a class that inherits from another class, we must place the name of the parent class inside round brackets in the definition of the child class. If a class is inheriting from another class, the __init__() method can be omitted altogether, but since we want to add some information to the child class, we define it. If we want to inherit certain variables of the parent class, we must call the __init__() method of the parent class too. This is done in lines 41 and 58, using the reserved super() class. This could also be done by writing Person.__init__() – note that in this case we don't use round brackets in the class name. Writing the super class name instead of super() can be useful if a class is inheriting from more than one classes.

Once we call the super().__init__() method, our child class will inherit everything that happens in the __init__() method of the parent class. If you run this script, you will get the Parent() class information of the Employee() object john.

The concept of classes in OOP is rather abstract and you might need some time to really grasp it. As a matter of fact, when using Pyo, it is rare that we need to define our own classes, as the available classes in Pyo cover a very wide range of applications. Still, we might find ourselves in a situation where our needs are not entirely covered by Pyo, in which case we might need to define a class of our own. In the next sections of this chapter, we will do exactly this.

8.2   Writing a Square Wave Oscillator Class with Duty Cycle Control

The first class we will define is a square wave oscillator with a settable duty cycle. The duty cycle is the same as the PWM (Pulse Width Modulation) technique we used in chapter 5 with the Arduino to fade an LED in and out. We define a value between 0 and 1 to represent the percentage of one cycle our oscillator will output a 1. For the remaining time of the cycle, it will output a -1. The result of controlling the duty cycle is a timbral change in the sound of the oscillator.

Ingredients:

- Python3
- The Pyo module
- The wxPython module, if you want to use Pyo's GUI
- A text editor (preferably an IDE with Python support)
- A terminal window in case your editor does not launch Python scripts

Process:

Script 8.5 includes the code of this class. It should be saved as a separate file, and loaded to another file where this oscillator will be used. When writing classes with PyoObjects, we need to inherit from the PyoObject() base class. Before anything else, we write a multi-line string, explaining what this class does, and providing a small example as to how it can be used. This is the documentation of our class, and we can call it using Python's built-in help(). By calling help(Square), this string will be printed on the console together with any string inherited from this class' super class.

Script 8.5 The square wave with settable duty cycle class.
```
1   from pyo import *
2
```

```
3   class Square(PyoObject):
4       """
5       A square wave oscillator with duty cycle control
6       :Parent: :py:class:'PyoObject'
7
8       :Args:
9           freq: float or PyoObject
10              Oscillator frequency
11          phase: float or PyoObject
12              Phase of the oscillator
13          duty: float or PyoObject
14              Duty cycle
15
16      >>> s = Server().boot()
17      >>> s.start()
18      >>> lfo = Sine(.25, mul=.5, add=.5)
19      >>> square = Square([200,202], duty=lfo, mul=.2).out()
20      """
21      def __init__(self,freq=1000,phase=0,duty=0.5,mul=1,add=0):
22          PyoObject.__init__(self, mul, add)
23          self._freq = freq
24          self._duty = duty
25          self._phase = phase
26          self._freq,self._phase,\
27          self._duty,mul,add,lmax=convertArgsToLists(
28                                      freq,phase,
29                                      duty,mul,add
30                                  )
31          self._phasor = Phasor(freq=self._freq,
32                                phase=self._phase)
33          self._comp = Compare(self._phasor, comp=self._duty,
34                               mul=2, add=-1)
35          self._sig = Sig(self._comp, mul=mul, add=add)
36          self._base_objs = self._sig.getBaseObjects()
37
38      def setFreq(self, freq):
39          self._freq = freq
40          self._phasor.setFreq(self._freq)
41
42      def setPhase(self, phase):
43          self._phase = phase
44          self._phasor.setPhase(self._phase)
45
46      def setDuty(self, duty):
47          self._duty = duty
48          self._comp.setComp(self._duty)
49
50      @property
51      def freq(self):
52          return self._freq
53
54      @freq.setter
55      def freq(self, freq):
```

```
56                  self.setFreq(freq)
57
58          @property
59          def phase(self):
60                  return self._phase
61
62          @phase.setter
63          def phase(self, phase):
64                  self.setPhase(phase)
65
66          @property
67          def duty(self):
68                  return self._duty
69
70          @duty.setter
71          def duty(self, duty):
72                  self.setDuty(duty)
73
74
75  if __name__ == "__main__":
76      s = Server().boot()
77      lfo = Sine(freq=.25, mul=.4, add=.5)
78      square = Square([200,202], duty=lfo, mul=.2).out()
79      sc = Scope(square, gain=1)
80      s.gui(locals())
```

In this class we will include five arguments, one for the frequency, one for the phase, one for the duty cycle, and the standard mul and add kwargs. In the parent class __init__() method, we only need to pass the mul and add kwargs. Once we do that, we don't need to do any other sort of handling for them, they will be handled by the PyoObject() parent class.

Two lines that need explaining are lines 26 to 30. These lines call the function convertArgsToLists() that comes with Pyo. This function is used to convert any non-list or non-PyoObjectBase() class arguments to lists. The PyoObjectBase() is a class that the PyoObject() class inherits from, so any class we write that inherits from the latter, will also inherit from the former. This means that if we pass any PyoObject, or an object of a class that we write that inherits from PyoObject(), then this function will leave this argument intact, otherwise, it will convert it to a list, if it is not one already. This function is used to allow multi-channel expansion, using Pyo's syntax, where we can expand the audio stream of a Pyo class to multiple channels by passing lists as arguments. The lmax is the maximum length of any possible list argument, among all the previous arguments in this line, that is returned by this function. In lines 31 to 35 we create out square wave oscillator. To output a waveform with sharp edges, but to also be able to control its duty cycle, we don't use Pyo's classes that create square waves, LFO() and RCOsc(), and we define this class in a different way than we did in chapter 3, where we used the Round() class. Here we use the Compare() class and pass the self._duty attribute of our class to its comp kwarg. Compare() takes a PyoObject as its first input, and compares its audio stream to its comp value. Depending on its mode, which defaults to "<", an object of this class outputs a 1 if the comparison is true, and a 0 if it is false. This way we can control the percentage of a cycle of our oscillator, by comparing the output of a Phasor(), which is a rising ramp from 0 to 1, to a value between 0 and 1. We set the `mul` attribute of the `self._comp` object to 2, and its `add` to -1, to bring its output to the full audio range. This class cannot be sent straight to the speakers, so we use a `Sig()` object as our output.

Line 36 defines a list of objects that is returned by the `getBaseObjects()` method of the `PyoObjectBase()` class. This list contains template Pyo audio stream objects, as many as the streams of this class we will create when initialising an object – which depends on whether we will pass lists to any of its kwargs. This is essentially the audio output that is seen by other Pyo objects, so we call it for the object we want to output, in this case, the `Sig()` object.

Once we are done with the `__init__()` method, we can define any other methods we want. In lines 38, 42, and 46, we define methods to override the first three kwargs of our class. The rest of the methods are different than anything else we have seen so far. These methods use what in Python is called a decorator. A decorator is defined with the @ symbol, and adds to the functionality of a method or function, depending on what the decorator we use does. We can write our own decorators, but Python contains built-in decorators too. A thorough explanation of decorators is beyond the scope of this chapter, so we will constrain this discussion to the specific decorators.

All Pyo objects use the built-in `property` decorator, which is the one we use here as well. It is a convenient way to get, set, and delete values in classes, without interfering immediately with variables that are treated as private. As soon as we decorate a method with @property, we can call a decorator with the name of the decorated method followed by the `setter` or `deleter` keywords, to create a method for setting, or deleting an attribute. The initial property is called the getter, because it gets the value of an attribute, and the other two are called the setter and deleter, respectively. In our case, we don't want to delete any attribute, so we omit the deleter.

Since the `self._freq`, `self._phase`, and `self._duty` attributes are initialised through kwargs with the `freq,` `phase`, and `duty` keywords, the user might try to get or set these values by typing the following line to set the frequency, or something equivalent for the other attributes.

```
square.freq = [250, 251]
```

By using this decorator, we permit the syntax above, without letting the user act upon the attributes we want to treat as private.

Another thing that needs to be discussed is the `if __name__ == "__main__"` test in line 75. This script is supposed to be loaded in other scripts so we can use this oscillator class together with more PyoObjects. When developing classes though, it is a good idea to write a chunk that will test the class. Remember from Chapter 2 how this test behaves. If this script is run, this test will be successful since the variable `__name__` will be assigned the string "`__main__`", but if this script is loaded in another script, then this variable will be assigned the name of the script that contains it, therefore, the `if` test in line 72 will fail and not run. By including this chunk in our script, we can run it to make sure our class behaves like we want it to.

The last thing to note is that even though we did not save the `mul` and `add` attributes to variables of the class, or we did not define an `out()`, `play()`, or `stop()` method, we can still use these effectively. This is because our class inherits from the `PyoObject()` class that includes these attributes and methods. If you write your own classes, you should always inherit from this base class to utilise Pyo to its full capacity.

Documentation Page:

Compare(): https://belangeo.github.io/pyo/api/classes/utils.html#compare

8.3   Writing a Triangle Wave Oscillator Class with Breakpoint Control

The next class we will write is a triangle wave oscillator with breakpoint control. This means that the waveform of the oscillator will be able to go from a backward sawtooth, through a triangle, to a forward sawtooth. As with the duty cycle control of the square wave oscillator, controlling the breakpoint of a triangle wave results in timbral changes.

Ingredients:

- Python3
- The Pyo module
- The wxPython module, if you want to use Pyo's GUI
- A text editor (preferably an IDE with Python support)
- A terminal window in case your editor does not launch Python scripts

Process:

Script 8.6 contains the code of this project. Most of the concepts are identical to the previous project, so no discussion on those is necessary. This code is copied from a GitHub repository and slightly modified so long lines fit the pages. This repository comes with a licence, this is the reason why the top comment is included. When sharing code, you will most likely want to include a licence. Whether you do or not, it is a good idea to include a comment at the beginning of your file, stating whether a licence is included, and if so, which licence is used. The GPL, or LGPL, along with other licences, like the BSD or MIT, enable free distribution of the code, even for commercial uses. Make sure to read a licence carefully before using it.

Script 8.6 The triangle wave with settable breakpoint class.

```
1   # Copyright 2017 Alexandros Drymonitis
2   #
3   # This code is based on the Pyo Python module
4   # and code by Olivier Belanger
5   # Pyo is released under the GNU GPL 3 Licence,
6   # so is this file
7   # A Licence copy should come with this code
8   # If not, please check <http://www.gnu.org/licenses/>
9   #
10  # This is an oscillator with a settable breakpoint
11
12  from pyo import *
13
14  class BrkPntOsc(PyoObject):
15      """
16      An oscillator with a settable breakpoint
17      resulting in a waveform that goes from a backward
18      sawtooth to a forward sawtooth.
```

```
19
20      :Parent: :py:class: `PyoObject'
21
22      :Args:
23
24          freq: float or PyoObject, optional
25              Frequency in cycles per second. Defaults to 100.
26          phase: float or PyoObject, optional
27              Phase of sampling,
28              expressed as a fraction of a cycle (0 to 1).
29              Defaults to 0.
30          brkpnt: float or PyoObject, optional
31              Point where the waveform breaks. From 0 to 1.
32              Defaults to 0.5.
33
34      >>> s = Server().boot()
35      >>> s.start()
36      >>> a = BrkPntOsc(200, brkpnt=.75, mul=.2).out()
37
38      """
39      def __init__(self, freq=100, phase=0,
40                  brkpnt=0.5, mul=1, add=0):
41          PyoObject.__init__(self, mul, add)
42          self._freq = freq
43          self._phase = phase
44          self._brkpnt = Sig(brkpnt)
45          self._invbrk = 1.0 - self._brkpnt
46          self._phasor=Phasor(freq=self._freq,
47                          phase=self._phase)
48          self._rising = (self._phasor/self._brkpnt) * \
49                          (self._phasor < self._brkpnt)
50          self._falling = (((self._phasor - self._brkpnt)/\
51                          self._invbrk) * (-1) + 1) * \
52                          (self._phasor >= self._brkpnt)
53          self._osc=Sig((self._rising+self._falling),
54                          mul=2,add=-1)
55          # A Sig is the best way to properly handle
56          # "mul" and "add" arguments.
57          self._output = Sig(self._osc, mul, add)
58          # Create the "_base_objs" attribute.
59          # This is the object's audio output.
60          self._base_objs = self._output.getBaseObjects()
61
62      def setFreq(self, x):
63          """
64          Replace the `freq' attribute.
65
66          :Args:
67
68          x: float or PyoObject
69              New `freq' attribute.
70
```

```
71              """
72              self._freq = x
73              self._phasor.freq = x
74
75      def setPhase(self, x):
76              """
77              Replace the 'phase' attribute.
78
79              :Args:
80
81              x: float or PyoObject
82                  New 'phase' attribute.
83
84              """
85              self._phase = x
86              self._phasor.phase = x
87
88      def setBrkPnt(self, x):
89              """
90              Replace the 'breakpoint' attribute.
91
92              :Args:
93
94              x: float or PyoObject
95                  New 'phase' attribute.
96
97              """
98              self._brkpnt.value = x
99
100     def play(self, dur=0, delay=0):
101             for key in self.__dict__.keys():
102                 if isinstance(self.__dict__[key], PyoObject):
103                     self.__dict__[key].play(dur, delay)
104             return PyoObject.play(self, dur, delay)
105
106     def stop(self):
107             for key in self.__dict__.keys():
108                 if isinstance(self.__dict__[key], PyoObject):
109                     self.__dict__[key].stop()
110             return PyoObject.stop(self)
111
112     def out(self, chnl=0, inc=1, dur=0, delay=0):
113             for key in self.__dict__.keys():
114                 if isinstance(self.__dict__[key], PyoObject):
115                     self.__dict__[key].play(dur, delay)
116             return PyoObject.out(self, chnl, inc, dur, delay)
117
118     @property
119     def freq(self):
120             """float or PyoObject.
121                 Fundamental frequency in cycles per second."""
```

```
122              return self._freq
123
124        @freq.setter
125              def freq(self, x): self.setFreq(x)
126
127        @property
128        def phase(self):
129              """float or PyoObject.
130                  Phase of sampling between 0 and 1."""
131              return self._phase
132
133        @phase.setter
134        def phase(self, x): self.setPhase(x)
135
136        @property
137        def breakpoint(self):
138              """float or PyoObject.
139                  Breakpoint of oscillator between 0 and 1."""
140              return self._brkpnt
141
142        @breakpoint.setter
143        def breakpoint(self, x): self.setBrkPnt(x)
144
145
146    if __name__ == "__main__":
147          # Test case . . .
148          s = Server().boot()
149
150          a = Sine(freq=.2, mul=.25, add=.5)
151          brk = BrkPntOsc(freq=200, brkpnt=a, mul=.2).out()
152
153          sc = Scope(brk)
154
155          s.gui(locals())
```

In addition to the multi-line string after the class definition, each method includes a string as well, stating what each of these does. These strings will also be printed to the console if we invoke the documentation of this class by typing `help(BrkPntOsc)`. Additionally, we can type `help(BrkPntOsc.setFreq)` to have only the documentation of this method printed. Other than that, the structure of this class is identical to that of the previous project, so what is left to discuss is how we achieve the control of the breakpoint. This happens in lines 44 to 54. In line 44 we set the breakpoint as the value of a `Sig()` object, and in line 45 we create a Pyo `ArithmeticDummy()` object with the inverse value of the breakpoint. Remember that when we apply arithmetic operations on PyoObjects, a dummy object is created that treats the result of the arithmetic operation as a Pyo audio stream.

In lines 48 and 50 we create the rising and falling part of the oscillator. The rising part is created by dividing the output of the `Phasor()` by the breakpoint, and letting that through only when the `Phasor()`'s output is less than the breakpoint. We achieve this sort of gating by multiplying the division by the result of the "less than" test in line 49. This will result in a ramp going from 0 to 1, for the percentage of the cycle set by the breakpoint.

Lines 50 to 52 work in a similar way, only we need to offset the output of `Phasor()` by the breakpoint value, so when the division is let through the gate created by the multiplication by the "greater than or equal to" test in line 52, it starts from zero instead of the breakpoint value. The multiplication by -1 and the addition of 1 change the direction of the ramp from upward to downward. Lines 50 to 52 result in a ramp going from 1 to 0 for the percentage of the cycle set by the inverse of the breakpoint.

In line 53 we create a `Sig()` object with the addition of the two ramps as its value. Since each ramp is gated alternatingly, their addition will result in an alternating upward and downward ramp, with their breakpoint being the value we pass through the class's `brkpnt` kwarg. We multiply this signal by 2 and subtract 1 to bring it to the full audio range, and then we create another `Sig()` object so we can utilise the `mul` and `add` attributes of the `PyoObject()` base class, as the corresponding kwargs of `self._osc` are already used.

Finally, in lines 100 to 116, the `play()`, `stop()`, and `out()` methods have been defined explicitly, the way they are defined in PyoObject class definitions. This was not done in the `Square()` class, which means that it is not necessary to include them in your class definitions, since they are inherited by the `PyoObject()` class. They are provided here as an example, in case you want to read them.

8.4   Conclusion

We have seen how to write our own classes in Python and some use cases where these can be useful. We have also seen the Pyo API and how to integrate this to our own classes, so we can have them interact with the rest of the Pyo classes. We briefly touched the decorator concept and saw how to create help documentation for our classes, by using multi-line strings in the class and each method definition. Even though Pyo is a complete toolkit that can cover a very wide range of applications with its native classes, we might find ourselves in a situation where we need to write our own class, for a certain functionality. Note that the classes we wrote in this chapter are pure Python, utilising Pyo classes for the audio calculations. Remember that Pyo is written in C, and only its interface is Python, as the latter alone does not meet the strict temporal constraints of audio. This means that for very advanced use cases, it might be necessary to write an external class in C. Writing external Pyo classes in C is beyond the scope of this book, but needing to do this will probably be a rarity, if not something you will never encounter.

8.5   Exercises

8.5.1   *Exercise 1*

Write documentation strings for all methods – including decorated methods – of the `Square()` class.

8.5.2   *Exercise 2*

Write an FM class that sets the modulator frequency explicitly, and not via a ratio. Both the carrier and the modulator oscillators should have a settable waveform, preferably with cross-fade. Make the cross-fade modulatable by being capable of being assigned a PyoObject and not only a float.

Tip: Use the waveforms you want for both the carrier and the modulator combined with `Selector()` to be able to cross-fade between them. The `LFO()` could be another choice, without a `Selector()`, but it does not cross-fade between its waveforms, and it does not have a sine wave, but only a modulated sine wave.

9 Switched on music21

In this chapter we will take a look at the music21 Python module, though we will only scratch the surface. We will also be approaching computer music from a more traditional angle. The two projects we will realise in this chapter are playing music by Bach using Pyo oscillators, and applying a simplified version of Schoenberg's twelve-tone system to create short compositions using Pyo.

What You Will Learn

After you have read this chapter you will:

- Know the basics of music21
- Know how to retrieve useful information from music21 in electronic music contexts
- Be able to apply more traditional music thinking into your Python scripts

---

9.1 What is music21?

Music21 is a toolkit aimed at computer-aided musicology, developed at the MIT (Cuthbert and Ariza, 2010). Despite not being aimed at computer-aided composition, music21 can provide a lot of useful and inspiring elements that can be used for the creation of algorithmic composition scripts. This Python module is very rich, as it contains numerous classes, and each class contains numerous methods, where each can be useful, depending on the context at hand. For our purposes, this module will provide useful tools to either produce compositions of our own, or play back composed acoustic music with electronic media, like oscillators. We will first create a script that applies a very simplistic version of Arnold Schoenberg's twelve-tone system to create algorithmic compositions, and then we will move on and create a script that will randomly choose a Bach piece from the music21 corpus, and will play it back with Pyo oscillators. Throughout this chapter, we will be using the music21 module, so go ahead and install it in your system by typing `pip install music21`.

9.2 Applying Schoenberg's Twelve-Tone System to Electronic Music

The twelve-tone system of Arnold Schoenberg (Haimo, 1993) is regarded as one of the most influential music revolutions of the 20th century. Initiated in the early 20th century, this system broke from the traditional approach in music composition and regarded all twelve tones of the

DOI: 10.4324/9781003386964-9

chromatic scale as equally important in a music piece. In this project we will utilise a function of the music21 module that creates a twelve by twelve matrix of twelve-tone series.

Ingredients:

- Python3
- The Pyo module
- The wxPython module, if you want to use Pyo's GUI
- The music21 module
- A text editor (preferably an IDE with Python support)
- A terminal window in case your editor does not launch Python scripts

Process:

To get a better idea of how we can utilise the music21 module for this project, let's first create a twelve by twelve matrix, where each row and each column will contain a twelve-tone series. In a terminal window, or using Jupyter, type the contents of Shell 9.1.

Shell 9.1 Creating a 12x12 twelve-tone matrix with music21.

```
>>> from music21 import *
>>> import random
>>> serie = [i for i in range(12)]
>>> random.shuffle(serie)
>>> matrix = serial.rowToMatrix(serie)
>>> print(matrix)
 0   5   1   8  10   2  11   3   7   6   9   4
 7   0   8   3   5   9   6  10   2   1   4  11
11   4   0   7   9   1  10   2   6   5   8   3
 4   9   5   0   2   6   3   7  11  10   1   8
 2   7   3  10   0   4   1   5   9   8  11   6
10   3  11   6   8   0   9   1   5   4   7   2
 1   6   2   9  11   3   0   4   8   7  10   5
 9   2  10   5   7  11   8   0   4   3   6   1
 5  10   6   1   3   7   4   8   0  11   2   9
 6  11   7   2   4   8   5   9   1   0   3  10
 3   8   4  11   1   5   2   6  10   9   0   7
 8   1   9   4   6  10   7  11   3   2   5   0
>>>
```

The `print(matrix)` line will print a matrix similar to the one in Shell 9.1. This matrix has twelve rows and twelve columns with twelve individual numbers each, from 0 to 11. The 0s of all rows and columns span across the matrix diagonally from left to right, top to bottom. This matrix is actually a string that separates row with newline characters. This is why when we print it with `print()` we get the separate rows like this. We will use one such matrix to produce a composition with a variable number of voices, but this number should be a value that produces a whole number when 12 is divided by it, so only the numbers 1, 2, 3, 4, 6, and 12 are possible.

We will divide this matrix by the number of voices and will assign the corresponding portion of the matrix to each voice. This is not how Schoenberg's twelve-tone system actually works, but for the purposes of this project, approaching the twelve-tone series this way is sufficient. We will also apply the class concept that was discussed in the previous chapter, as we will create classes for the oscillators and sequencers. We will break the code into small chunks and discuss as we go.

The first chunk is shown in Script 9.1. This is the top part of the full code where we import all the modules we will need and we assign some macros and global variables. Notice that we don't use the asterisk to import everything either from Pyo or from music21. This is because there is a chance we will run into name clashes with certain classes of the two modules, so we load them without "polluting our namespace". The macros concern the lowest note as a MIDI number, the number of set voices as number of oscillators, the numerator and denominator of the time division – in this case four quarters – the minimum note duration – a sixteenth note – and the BPM.

Lines 14 and 21 might ring a bell from Chapter 7. We will use the same FM3() class we used in the regression Neural Network, so we will use part of the code of that project. Some parameters though will not be hard-coded, so we include only the two ratio values in the `param_list`. In line 16 we create our metronome. We will use a technique similar to the drum machine we used in Chapters 3 and 6, so we need a `Metro()` object to drive our sequencer. Lines 17 and 18 create a `Counter()` and `Select()` object respectively, so we can stop the metronome once the entire twelve-tone matrix has been played. We will see later on how these are used.

Script 9.1 Importing modules and setting global variables.

```
 1  import music21 as m21
 2  import pyo
 3  import random
 4
 5  BASENOTE = 48
 6  NUMOSC = 2
 7  METER = 4
 8  BASEDUR = 4
 9  MINDUR = 16
10  BPM = 80
11
12  s = pyo.Server().boot()
13
14  t = pyo.HarmTable([1, 0.1])
15
16  metro = pyo.Metro(time=pyo.beatToDur(1/int(MINDUR/4),BPM))
17  pyocounter = pyo.Counter(metro)
18  select = pyo.Select(pyocounter)
19
20  param_lists = [[0.33001, 2.9993],
21                 [0.33003, 2.9992]]
22
```

For the next chunk we will move a bit further and go straight to the `if __name__ == "__main__"` part of the code. This is shown in Script 9.2. The first part is similar to Shell 9.1, so we will start the discussion below that.

Script 9.2 The if __name__ == "__main__" chunk.

```
78  if __name__ == "__main__":
79      # create a random twelve tone line
80      serie = [i for i in range(12)]
```

```
81      random.shuffle(serie)
82      # create a string with twelve lines
83      # with unique twelve tone lines
84      # separated with \n
85      series = m21.serial.rowToMatrix(serie)
86      # separate the string above to twelve strings
87      twelvetone_list = series.split('\n')
88      # get a list with twelve lists with integers
89      # from the strings above
90      matrix = []
91      counter = 0
92      for line in twelvetone_list:
93          matrix.append([])
94          splitline = line.split(' ')
95          for string in splitline:
96              if len(string) > 0:
97                  matrix[counter].append(int(string))
98          counter += 1
99      # flatten the 2D list to one 1D list
100     matrix_flat = []
101     for serie in matrix:
102         for note in serie:
103             matrix_flat.append(note)
104
105     listlen = int(len(matrix_flat)/NUMOSC)
106     notelist = [matrix_flat[i*listlen:(i+1)*listlen]
107                 for i in range(NUMOSC)]
108
109     durlist = []
110     masklist = []
111     for i in range(NUMOSC):
112         masklist.append([])
113         durlist.append([])
114         for j in range(int(len(notelist[i])/METER)):
115             masklocal = rand_durs(METER, MINDUR)
116             for dur in masklocal:
117                 masklist[i].append(1)
118                 for k in range(dur-1):
119                     masklist[i].append(0)
120                 durlist[i].append(dur)
121
```

In line 87 we split the matrix string based on the newline character, so we end up with a list of single strings, where each string contains twelve unique values. Then in line 92 we run a loop and split each line based on the white space, and then detect strings that have at least one character, convert them to integers, and store them to the matrix list. We then flatten the 2D matrix to a 1D list, to more easily break it to as many chunks as we like. Once we have a flat matrix, we chop it up to as many pieces as the NUMOSC macro, using list comprehension. This last list, notelist, is a list of lists where each sublist contains all the notes in MIDI numbers for one voice. We will be iterating over this list with our sequencing mechanism to pass pitches to our oscillators.

Line 109 onward contains the code where we set the durations for the oscillators, and the masking lists for the triggers. To simplify things, we will create bars of four notes each, hence the division by METER which equals 4, in line 114. In line 115 we call a function called `rand_durs()` that takes the number of random values to be produced, and the sum these values should produce. In this case, we want to create four random values for one bar, that sum up to MINDUR, which equals 16. This means that we will be using the sixteenth note as the shortest duration, but that can easily change by changing the MINDUR macro in line 9 in Script 9.1. This function is shown in Script 9.3, and it is copied from StackOverflow.[1]

Script 9.3 A function that produces a certain number of random values with a specific sum.
```
72 def rand_durs(n, total):
73     dividers = sorted(random.sample(range(1, total), n - 1))
74     return [a - b for a, b in zip(dividers + \
75             [total], [0] + dividers)]
76
```

By calling the function of Script 9.3, we can create a variety of rhythmic patterns, but always have four notes that fit to a 4/4 bar. In line 116 of Script 9.2, we run a loop to store 1s and 0s to our masking list. We take for granted that the first beat will always have a note, as we don't use rests in this project, so line 117 appends a 1 and then we move on to append as many 0s as the actual duration minus 1. If for example we get a random value of 4, which is a quarter note, we will store a 1 in the first position, and three 0s after that. This way, our sequencer will trigger the oscillator in the first step, but it will be gated for the next three, so the oscillator will sound for the duration of one quarter, while our sequencer will tick at the tempo of sixteenths.

Let us now move to the two class definitions. The first one, the class we used in Chapter 7, is shown in Script 9.4. As we did not discuss this class in Chapter 7, we will briefly discuss it here. This class creates three stacked sine wave oscillators, where one modulates the frequency of the next. The first two, the two modulators, are created with the `Sine()` class, while the carrier is a table lookup oscillator, created with the `Osc()` class. The waveform this oscillator looks up is created in line 14 in Script 9.1, with the `HarmTable()` class. This version of this class has no methods as we don't need to set the frequency or any other attribute explicitly. We will use this class as the sound generator in the next class we will define, which will include all the sequencing objects we need.

Script 9.4 The FM3 class.
```
23 class FM3(pyo.PyoObject):
24     def __init__(self,fcar,ratio1,ratio2,mul=1,add=0):
25         pyo.PyoObject.__init__(self, mul, add)
26         fcar,ratio1,ratio2,\
27         mul,add,lmax=pyo.convertArgsToLists(
28                         fcar,ratio1,
29                         ratio2,mul,add
30                     )
31         self._fcar = pyo.Sig(fcar)
32         self._fmod = pyo.SigTo(ratio1, mul=self._fcar)
33         self._fmodmod = pyo.SigTo(ratio2, mul=self._fmod)
34         self._amod = pyo.SigTo(8, mul=self._fmod)
35         self._amodmod = pyo.SigTo(4, mul=self._fmodmod)
```

```
36          self._modmod=pyo.Sine(self._fmodmod,
37                              mul=self._amodmod)
38          self._mod=pyo.Sine(self._fmod+self._modmod,
39                              mul=self._amod)
40          self._car = pyo.Osc(t, fcar+self._mod, mul=0.2)
41          self._eq=pyo.EQ(self._car,freq=fcar,
42                          q=0.707,boost=-12)
43          self._out = pyo.DCBlock(self._eq, mul=mul)
44          self._base_objs = self._out.getBaseObjects()
45
```

Additionally, this version has two differences from the one in Chapter 7. The conversion of non-list arguments to lists, in lines 26 to 30, which we have seen in the previous chapter, and the call to the getBaseObjects() method in line 44. In Chapter 7 we didn't need to call this function because our class did not interact with any other PyoObjects. In this project though, we need to call this function because it returns a list of objects that are seen by other PyoObjects. If we want our class to interact in any way with other Pyo classes, we must call this function for the PyoObject we want to use as our class's output.

Scrip 9.5 shows the code of the second class we define in this project. This is the class where we use a technique similar to the one we used in the drum machine project in Chapters 3 and 6. The arguments we need in this class are the ID number of the object, and the notes, masking, and durations lists. We also need the ratio items from the param_lists list, because we will be creating an FM3() object in here. Line 52 creates an offset for the notes in the notelist based on the ID. This is done so the voices of our composition will not play in the same octave, rendering them distinguishable, since their timbres will be similar.

Script 9.5 The Schoenberg sequencing class.

```
46 class Schoenberg(pyo.PyoObject):
47     def __init__(self,id=0,notelist=None,
48                  masklist=None,durlist=None,mul=1,add=0):
49         pyo.PyoObject.__init__(self, mul, add)
50         self._ratio1 = param_lists[id%2][0]
51         self._ratio2 = param_lists[id%2][1]
52         self._note_offset = id*12
53         self._mask = pyo.Iter(metro, choice=masklist)
54         self._beat = metro * self._mask
55         self._notes = pyo.Iter(self._beat, choice=notelist)
56         self._durs = pyo.Iter(self._beat, choice=durlist)
57         self._adsr=pyo.Adsr(dur=pyo.beatToDur(1/MINDUR,BPM),
58                             mul=(1/NUMOSC))
59         self._tf = pyo.TrigFunc(self._beat, self.adsr_ctl)
60         self._fm = FM3(pyo.MToF(self._notes+\
61                         (BASENOTE+self._note_offset)),
62                         self._ratio1,self._ratio2,
63                         mul=self._adsr)
64         self._base_objs = self._fm.getBaseObjects()
65
66     def adsr_ctl(self):
67         self._adsr.setDur(pyo.beatToDur(self._durs.get()/\
68                           (MINDUR/4),BPM))
```

```
69              self._adsr.play()
70
71
```

In lines 53 and 54 we create the local metronome that will trigger each object based on its masking list. This is done in a similar way as with the drum machine project in Chapters 3 and 6. Lines 55 and 56 initialise two Iter() objects that are triggered by the local metronome, one for the notes and one for the durations. Finally, we create an ADSR envelope, a Trig-Func() that will set the duration of the envelope and will trigger it, and the FM3() object that will produce the sound. The notes of this object are derived from the _notes object of the Iter() class, passed to an MToF() function, after the BASENOTE and ID-based offset have been added. This is the object we want to output for the rest of the Pyo ecosystem to interact with, so we call the getBaseObjects() method for this object.

It is possible to call the getBaseObjects() method for the _fm object in the Sch-oenberg() class, because we called this function in Script 9.4, for the _out object, in line 44. If we omitted calling this function in the FM3() class, we would not be able to do this in the Schoenberg() class, as the FM3() class would not be able to interact with the rest of the Pyo ecosystem. Note also that we did not call the convertArgsToLists() function, because we will be creating single-stream objects of this class. The reason why is explained later on.

The last bit of this class is the adsr_ctl() function. This function sets the duration of the ADSR envelope and triggers it. The duration is derived from Pyo's beatToDur() function. Since we want to have an easy interface to set the minimum duration, by changing one macro only, we need to divide the stored duration values by the minimum duration divided by 4, to get the value we need. If, for example, we get a duration of 1, which is a quarter note, the output of beatToDur() will be $1/(1/4) = 1/0.25 = 4$. This means that we will get the duration of four sixteenth notes for the tempo set by the BPM macro, which is what we want.

There is one last part left before we finish the code for this project. This is shown in Script 9.6. This last part comes right after Script 9.2, and it is included in the if __name__ == "__main__" chunk, hence the indentation. In this bit, we set the maximum count for the beat counter, in line 123. This is the number of any sublist in the masklist, as this list contains a list for each voice with 1s and 0s, for every beat of the metronome. We also set the value for the Select() object so it can send a trigger when pyocounter reaches its maximum value – counting is zero-based, so the value for select is pyocounter's maximum value minus 1.

Script 9.6 The Schoenberg object and the beat counter settings.

```
122         # set the maximum count based on the mask list
123         pyocounter.setMax(len(masklist[0]))
124         select.setValue(len(masklist[0])-1)
125
126         schoenberg = [Schoenberg(i,notelist[i],masklist[i],
127                         durlist[i]) for i in range(NUMOSC)]
128         mix = pyo.Mix(schoenberg, voices=2).out()
129
130         tf = pyo.TrigFunc(select, metro.stop)
131
132         metro.play()
133
134         s.gui(locals())
```

We then create a list with `Schoenberg()` objects and we pass all the necessary arguments. We could have created the `schoenberg` object without the list comprehension, just by passing the entire lists of lists, as in the following line.

```
schoenberg = Schoenberg(notelist,masklist,durlist)
```

If we created this object this way though, it would not be easy to set the `id` attribute which we need, so we can set an offset for the frequencies. This is because when calling the `getBase-Objects()` method, our class is treated as the class for which we call this function. In our case, since in the `Schoenberg()` class we call this function for the FM3() class, which in turn calls this function for a `DCBlock()` object, the `schoenberg` object is treated as an object of the `DCBlock()` class. Even if we defined a method to set the `id` attribute explicitly, we would get an `AttributeError` saying that the `DCBlock()` object has no such attribute.

After we create our list with `Schoenberg()` objects, we input it into a `Mix()` object, for cleaner coding, and send it to the speakers. We then create a `TrigFunc()` object that will stop the metronome when it receives a trigger signal from the select object. This will happen when the `pyocounter` reaches its maximum value. This way we will iterate over the algorithmic composition only once, and we will not loop through it. Script 9.6 concludes the code for this project. The full code is shown in Script 9.7.

Script 9.7 The full Schoenberg style code.
```
 1 import music21 as m21
 2 import pyo
 3 import random
 4
 5 BASENOTE = 48
 6 NUMOSC = 2
 7 METER = 4
 8 BASEDUR = 4
 9 MINDUR = 16
10 BPM = 80
11
12 s = pyo.Server().boot()
13
14 t = pyo.HarmTable([1, 0.1])
15
16 metro = pyo.Metro(time=pyo.beatToDur(1/int(MINDUR/4),BPM))
17 pyocounter = pyo.Counter(metro)
18 select = pyo.Select(pyocounter)
19
20 param_lists = [[0.33001, 2.9993],
21                [0.33003, 2.9992]]
22
23 class FM3(pyo.PyoObject):
24     def __init__(self,fcar,ratio1,ratio2,mul=1,add=0):
25         pyo.PyoObject.__init__(self, mul, add)
26         fcar,ratio1,ratio2,\
27         mul,add,lmax=pyo.convertArgsToLists(
28                          fcar,ratio1,
29                          ratio2,mul,add
30                      )
```

```
31          self._fcar = pyo.Sig(fcar)
32          self._fmod = pyo.SigTo(ratio1, mul=self._fcar)
33          self._fmodmod = pyo.SigTo(ratio2, mul=self._fmod)
34          self._amod = pyo.SigTo(8, mul=self._fmod)
35          self._amodmod = pyo.SigTo(4, mul=self._fmodmod)
36          self._modmod = pyo.Sine(self._fmodmod,
37                                  mul=self._amodmod)
38          self._mod=pyo.Sine(self._fmod+self._modmod,
39                             mul=self._amod)
40          self._car = pyo.Osc(t, fcar+self._mod, mul=0.2)
41          self._eq=pyo.EQ(self._car, freq=fcar,
42                          q=0.707, boost=-12)
43          self._out = pyo.DCBlock(self._eq, mul=mul)
44          self._base_objs = self._out.getBaseObjects()
45
46  class Schoenberg(pyo.PyoObject):
47      def __init__(self,id=0,notelist=None,
48                   masklist=None,durlist=None,mul=1,add=0):
49          pyo.PyoObject.__init__(self, mul, add)
50          self._ratio1 = param_lists[id%2][0]
51          self._ratio2 = param_lists[id%2][1]
52          self._note_offset = id*12
53          self._mask = pyo.Iter(metro, choice=masklist)
54          self._beat = metro * self._mask
55          self._notes = pyo.Iter(self._beat, choice=notelist)
56          self._durs = pyo.Iter(self._beat, choice=durlist)
57          self._adsr=pyo.Adsr(dur=pyo.beatToDur(1/MINDUR,BPM),
58                              mul=(1/NUMOSC))
59          self._tf = pyo.TrigFunc(self._beat, self.adsr_ctl)
60          self._fm = FM3(pyo.MToF(self._notes+\
61                         (BASENOTE+self._note_offset)),
62                         self._ratio1,self._ratio2,
63                         mul=self._adsr)
64          self._base_objs = self._fm.getBaseObjects()
65
66      def adsr_ctl(self):
67          self._adsr.setDur(pyo.beatToDur(self._durs.get()/\
68                            (MINDUR/4),BPM))
69          self._adsr.play()
70
71
72  def rand_durs(n, total):
73      dividers = sorted(random.sample(range(1, total),n - 1))
74      return [a - b for a, b in zip(dividers + \
75             [total], [0] + dividers)]
76
77
78  if __name__ == "__main__":
79      # create a random twelve tone line
80      serie = [i for i in range(12)]
81      random.shuffle(serie)
82      # create a string with twelve lines
```

```
83        # with unique twelve tone lines
84        # separated with \n
85        series = m21.serial.rowToMatrix(serie)
86        # separate the string above to twelve strings
87        twelvetone_list = series.split('\n')
88        # get a list with twelve lists with integers
89        # from the strings above
90        matrix = []
91        counter = 0
92        for line in twelvetone_list:
93            matrix.append([])
94            splitline = line.split(' ')
95            for string in splitline:
96                if len(string) > 0:
97                    matrix[counter].append(int(string))
98            counter += 1
99        # flatten the 2D list to one 1D list
100       matrix_flat = []
101       for serie in matrix:
102           for note in serie:
103               matrix_flat.append(note)
104
105       listlen = int(len(matrix_flat)/NUMOSC)
106       notelist = [matrix_flat[i*listlen:(i+1)*listlen]
107                   for i in range(NUMOSC)]
108
109       durlist = []
110       masklist = []
111       for i in range(NUMOSC):
112           masklist.append([])
113           durlist.append([])
114           for j in range(int(len(notelist[i])/METER)):
115               masklocal = rand_durs(METER, MINDUR)
116               for dur in masklocal:
117                   masklist[i].append(1)
118                   for k in range(dur-1):
119                       masklist[i].append(0)
120                   durlist[i].append(dur)
121
122       # set the maximum count based on the mask list
123       pyocounter.setMax(len(masklist[0]))
124       select.setValue(len(masklist[0])-1)
125
126       schoenberg = [Schoenberg(i,notelist[i],masklist[i],
127                     durlist[i]) for i in range(NUMOSC)]
128       mix = pyo.Mix(schoenberg, voices=2).out()
129
130       tf = pyo.TrigFunc(select, metro.stop)
131
132       metro.play()
133
134       s.gui(locals())
```

Documentation Page:

`Select()`: https://belangeo.github.io/pyo/api/classes/triggers.html#select

9.2.1   *Displaying the Score*

One nice feature of the music21 module is that it can create scores in the MusicXML format. To be able to see scores, you must have either Finale, Sibelius, or MuseScore installed in your system. Shell 9.2 is copied from music21's documentation, and displays the score of a Bach piece, retrieved from the corpus of music21. If you have one of these score engraving software installed in your system, you should see a score like the one in Figure 9.1. If you can't see the score, head to music21's documentation.[2] With music21, though, we can create our own scores too, so we will utilise this feature and create a score for our algorithmic composition.

Shell 9.2 Display a Bach score with music21.

```
>>> from music21 import *
>>> b = corpus.parse('bach/bwv66.6')
>>> b.show()
```

Ingredients:

- Python3
- The Pyo module
- The wxPython module, if you want to use Pyo's GUI
- The music21 module
- Finale, Sibelius, or MuseScore
- A text editor (preferably an IDE with Python support)
- A terminal window in case your editor does not launch Python scripts

Figure 9.1 A Bach score displayed with MuseScore.

Process:

We will insert some functions into the code of Script 9.7 that will convert the notes and durations data to information for music21 objects. The first thing we will look at is the part in the i f __name__ == "__main__" chunk, where we will create the score and the parts, and we will insert the latter to the former. Script 9.8 shows these lines that should go after the initialisation of the schoenberg and Mix() objects.

Script 9.8 Creating the score and parts.
```
237        score = m21.stream.Score(id="Schoenberg Style Piece")
238        parts = store_parts(notelist, durlist)
239
240        for i in range(NUMOSC):
241            score.insert(0, parts[i])
```

In line 238 we call the store_parts() function that we define in Script 9.9. This function creates a music21 stream.Part() object for each voice, and a list of lists with music21 stream.Measure() objects, one for each part. The latter stores music21 note.Note() objects, one for each note in the notelist. The notes are named after the Dutch note names – defined in lines 144 and 145 in Script 9.9 – with the octave number appended to the note name. The two voices of this project use their ID to set a note offset, so they don't play at the same octave. When displaying a score though, we want the staff with the higher notes to be displayed on top, hence the index inversion in line 153.

Script 9.9 The store_parts() function.
```
143 def store_parts(notelist, durlist):
144      notes = ["C","C#","D","D#","E","F",
145                "F#","G","G#","A","B-","B"]
146      measures = []
147      parts = []
148      for i in range(NUMOSC):
149          parts.append(m21.stream.Part(id="part"+str(i)))
150          measures.append([])
151          # invert note lists to get higher notes
152          # in the top staff
153          invndx = NUMOSC - 1 - i
154          # every twelve tone line is divided
155          # to three bars of four notes
156          for j in range(int(len(notelist[invndx])/4)):
157              measures[i].append(
158                      m21.stream.Measure(number=j+1)
159                  )
160          ts = m21.meter.TimeSignature('4/4')
161          measures[i][0].insert(0, ts)
162          for k in range(4):
163              # create a string with the note name
164              # and the octave number
165              octave = str(invndx+int(BASENOTE/12))
166              notestr = notes[notelist[invndx][(j*4)+k]]+\
167                  octave
```

```
168                        thisdur = durlist[invndx][(j*4)+k]
169                        notedata = get_notedata(thisdur)
170                        for ndx, noteinfo in enumerate(notedata):
171                            note=m21.note.Note(notestr,
172                                               type=noteinfo[0],
173                                               dots=noteinfo[1])
174                        if ndx == 0 and len(notedata) > 1:
175                            note.tie = m21.tie.Tie("start")
176                        elif ndx == (len(notedata)-1) and \
177                            len(notedata) > 1:
178                            note.tie = m21.tie.Tie("stop")
179                        measures[i][j].append(note)
180            measures[i][len(measures[i])-1].rightBarline='final'
181            parts[i].append(measures[i])
182     return parts
```

In line 160 we define the time signature as 4/4 and insert that at the beginning of each measure. We then run a loop to store all the notes. In line 169 we call the `get_notedata()` function that returns note types as strings – "eighth", "quarter", etc. – and the number of dots, in case a note is dotted. This function is shown in Script 9.10. Once we have this information, we can create a `note.Note()` object with the correct name, note type, and number of dots.

Since the algorithm creates notes of various durations, it might be that we need to tie notes. For example, with a base duration of a sixteenth note, if we get a value of 5, it means that we have a quarter note tied to a sixteenth note. In this case, `get_notedata()` will return a list of lists, one for each note in the tied group. In line 174, we test if we are in the first item of this list, and if there are more than one items, which means we must start a tie. In line 176 we test if we are in the last item of the list and if the list has more than one items, which means we must stop the tie. We then append the `note.Note()` object to the measure of the current voice.

When this loop is done, we set the right bar line to a final one, so the score displays a double bar line, with the right one being bold, to indicate the end of the score. Then we append the list of measures to the part of the voice we are processing. When we are done with all the parts, we `return` the list of `stream.Parts()` objects.

The `get_notedata()` function, shown in Script 9.10, is a bit more complicated, because we need to handle various scenarios of note durations. As mentioned above, we might need to dot a note, once or twice, and we might also need to tie it to another note, which in turn might need to be dotted too. The best way to handle this is with a recursive function, like this one.

Script 9.10 The get_notedata() function.

```
94 def get_notedata(thisdur):
95      notetypes = {1:"16th",2:"eighth",4:"quarter",
96                   8:"half",16:"whole"}
97      basedur = 0
98      dots = 0
99      notedata = []
100     if thisdur in notetypes.keys():
101         notedata.append([notetypes[thisdur], dots])
102     else:
103         for dur in notetypes.keys():
104             if dur > thisdur:
105                 basedur = dur
```

```
106                     break
107             halfway = int(basedur-(basedur/4))
108             # look for a dotted note
109             #only if there is such a possibility
110             if thisdur == halfway:
111                 notedata.append([notetypes[int(basedur/2)], 1])
112             elif thisdur > halfway:
113                 mindur = 0
114                 dots = get_num_dots(basedur, thisdur)
115                 # if we didn't catch thisdur we don't have dots
116                 while dots == 0:
117                     # so we decrement and test again
118                     thisdur -= 1
119                     # and increment the number of sixteenth notes
120                     mindur += 1
121                     dots = get_num_dots(basedur, thisdur)
122                 notedata.append([notetypes[int(basedur/2)],
123                                     dots])
124                 # accumulate sixteenths
125                 # to assemble longer durations
126                 if mindur > 0:
127                     if mindur > 1:
128                         # recurse until we exhaust our notes
129                         notedata.append(get_notedata(mindur)[0])
130                     else:
131                         notedata.append([notetypes[1], 0])
132             # if we don't have a dotted note
133             # we must look for ties
134             else:
135                 # first tied note is half the base dur
136                 notedata.append([notetypes[int(basedur/2)], 0])
137                 # get the remainder of the duration
138                 thisdur -= int(basedur/2)
139                 notedata.append(get_notedata(thisdur)[0])
140     return notedata
```

The first thing we need to do is to check if a note duration is included in the notetypes dictionary. In this case, we just append the value of the key set by the note duration, and 0 number of dots. If this is not the case, we must check if this note is dotted. This is true if the duration is equal to or greater than its duration plus half its duration. We set the first dictionary key that is greater than the note duration as a base, and derive the half way between this duration and the previous, by subtracting one quarter of our base duration from itself, in line 107. Then we check for equality, which means our note is single dotted, or if the note duration is greater than that, which means that our note is both dotted and tied.

In line 114 we call the get_num_dots() function which will be shown in the next script. This function returns the number of dots only if the queried duration is not tied, otherwise it returns a 0. If this is the case, we run the while loop in line 116, every time decrementing the note duration by one, until we get the number of dots we need. Every time we decrement the note duration, we increment a variable that holds the number of notes with the minimum note duration that will be tied to the current note.

When we exit the loop, we check if we do have tied notes by querying the value of mindur. If it is greater than 1, we recursively call the get_notedata() function, and append its output, a process which will continue until all the dots and tied notes, and their dots, have been exhausted. If mindur is 1, it means that we must tie a single note of the minimum duration, without a dot.

If the note duration is less than halfway between itself and the previous duration, we first store the duration type that is immediately less than the current duration – half the base duration, which is actually the immediately greater one – and then we recursively call the get_notedata() function again until we exhaust all the notes.

The last function we need to discuss is get_num_dots(), shown in Script 9.11. This function applies a similar technique to the get_notedata() function, as it checks for equality between the two durations, every time halving one duration, until it reaches 1 – by checking if there are any decimals in the value, in line 83 – or until it reaches equality. If equality is not reached, the function returns 0, otherwise it returns the number of iterations in the while loop.

Script 9.11 The get_num_dots() function.
```
78 def get_num_dots(dur1, dur2):
79     counter = 0
80     dots = 0
81     halfway = int(dur1-(dur1/4))
82     # while there are no decimals in dur1
83     while (dur1 % 2) == 0:
84         # keep on halving it
85         dur1 /= 2
86         counter += 1
87         # if it equals dur2, it means the note is dotted
88         if halfway == dur2:
89             dots = counter
90             break
91     return dots
```

These three functions were discussed in reverse order compared to the order they are written in the full script. The full code is shown in Script 9.12. Note that since we display the score, we don't call Pyo's GUI, so we start the server by calling s.start() in line 243. In line 247 we call score.show() which will keep the script alive, much like the s.gui(locals()) does. We also delay starting the metronome by five seconds, with CallAfter(), to give some time to the score rendering program to start. This is done in line 244.

Script 9.12 The full code of the Schoenberg style algorithmic composition with the sheet music.
```
1 import music21 as m21
2 import pyo
3 import random
4
5 BASENOTE = 48
6 NUMOSC = 2
7 METER = 4
8 BASEDUR = 4
9 MINDUR = 16
10 BPM = 80
```

```
11
12 s = pyo.Server().boot()
13
14 t = pyo.HarmTable([1, 0.1])
15
16 metro = pyo.Metro(time=pyo.beatToDur(1/int(MINDUR/4),BPM))
17 pyocounter = pyo.Counter(metro)
18 select = pyo.Select(pyocounter)
19
20 param_lists = [[0.33001, 2.9993],
21                 [0.33003, 2.9992]]
22
23 class FM3(pyo.PyoObject):
24     def __init__(self,fcar,ratio1,ratio2,mul=1,add=0):
25         pyo.PyoObject.__init__(self, mul, add)
26         fcar,ratio1,ratio2,\
27         mul,add,lmax=pyo.convertArgsToLists(
28                             fcar,ratio1,
29                             ratio2,mul,add
30                         )
31         self._fcar = pyo.Sig(fcar)
32         self._fmod = pyo.SigTo(ratio1, mul=self._fcar)
33         self._fmodmod = pyo.SigTo(ratio2, mul=self._fmod)
34         self._amod = pyo.SigTo(8, mul=self._fmod)
35         self._amodmod = pyo.SigTo(4, mul=self._fmodmod)
36         self._modmod = pyo.Sine(self._fmodmod,
37                                 mul=self._amodmod)
38         self._mod=pyo.Sine(self._fmod+self._modmod,
39                                 mul=self._amod)
40         self._car = pyo.Osc(t, fcar+self._mod, mul=0.2)
41         self._eq=pyo.EQ(self._car, freq=fcar,
42                                 q=0.707, boost=-12)
43         self._out = pyo.DCBlock(self._eq, mul=mul)
44         self._base_objs = self._out.getBaseObjects()
45
46 class Schoenberg(pyo.PyoObject):
47     def __init__(self,id=0,notelist=None,
48                     masklist=None,durlist=None,mul=1,add=0):
49         pyo.PyoObject.__init__(self, mul, add)
50         self._ratio1 = param_lists[id%2][0]
51         self._ratio2 = param_lists[id%2][1]
52         self._note_offset = id*12
53         self._mask = pyo.Iter(metro, choice=masklist)
54         self._beat = metro * self._mask
55         self._notes = pyo.Iter(self._beat, choice=notelist)
56         self._durs = pyo.Iter(self._beat, choice=durlist)
57         self._adsr=pyo.Adsr(dur=pyo.beatToDur(1/MINDUR,BPM),
58                                 mul=(1/NUMOSC))
59         self._tf = pyo.TrigFunc(self._beat, self.adsr_ctl)
60         self._fm = FM3(pyo.MtoF(self._notes+\
61                     (BASENOTE+self._note_offset)),
62                         self._ratio1,self._ratio2,
```

```
63                          mul=self._adsr)
64          self._base_objs = self._fm.getBaseObjects()
65
66      def adsr_ctl(self):
67          self._adsr.setDur(pyo.beatToDur(self._durs.get()/\
68                          (MINDUR/4),BPM))
69          self._adsr.play()
70
71
72  def rand_durs(n, total):
73      dividers = sorted(random.sample(range(1, total),n - 1))
74      return [a - b for a, b in zip(dividers + \
75              [total], [0] + dividers)]
76
77
78  def get_num_dots(dur1, dur2):
79      counter = 0
80      dots = 0
81      halfway = int(dur1-(dur1/4))
82      # while there are no decimals in dur1
83      while (dur1 % 2) == 0:
84          # keep on halving it
85          dur1 /= 2
86          counter += 1
87          # if it equals dur2, it means the note is dotted
88          if halfway == dur2:
89              dots = counter
90              break
91      return dots
92
93
94  def get_notedata(thisdur):
95      notetypes = {1:"16th",2:"eighth",4:"quarter",
96                      8:"half",16:"whole"}
97      basedur = 0
98      dots = 0
99      notedata = []
100     if thisdur in notetypes.keys():
101         notedata.append([notetypes[thisdur], dots])
102     else:
103         for dur in notetypes.keys():
104             if dur > thisdur:
105                 basedur = dur
106                 break
107         halfway = int(basedur-(basedur/4))
108         # look for a dotted note
109         #only if there is such a possibility
110         if thisdur == halfway:
111             notedata.append([notetypes[int(basedur/2)], 1])
112         elif thisdur > halfway:
113             mindur = 0
114             dots = get_num_dots(basedur, thisdur)
```

```
115              # if we didn't catch thisdur we don't have dots
116              while dots == 0:
117                  # so we decrement and test again
118                  thisdur -= 1
119                  # and increment the number of sixteenth notes
120                  mindur += 1
121                  dots = get_num_dots(basedur, thisdur)
122              notedata.append([notetypes[int(basedur/2)],
123                                  dots])
124          # accumulate sixteenths
125          # to assemble longer durations
126          if mindur > 0:
127              if mindur > 1:
128                  # recurse until we exhaust our notes
129                  notedata.append(get_notedata(mindur)[0])
130              else:
131                  notedata.append([notetypes[1], 0])
132      # if we don't have a dotted note
133      # we must look for ties
134      else:
135          # first tied note is half the base dur
136          notedata.append([notetypes[int(basedur/2)], 0])
137          # get the remainder of the duration
138          thisdur -= int(basedur/2)
139          notedata.append(get_notedata(thisdur)[0])
140      return notedata
141
142
143 def store_parts(notelist, durlist):
144     notes = ["C","C#","D","D#","E","F",
145             "F#","G","G#","A","B-","B"]
146     measures = []
147     parts = []
148     for i in range(NUMOSC):
149         parts.append(m21.stream.Part(id="part"+str(i)))
150         measures.append([])
151         # invert note lists to get higher notes
152         # in the top staff
153         invndx = NUMOSC - 1 - i
154         # every twelve tone line is divided
155         # to three bars of four notes
156         for j in range(int(len(notelist[invndx])/4)):
157             measures[i].append(
158                         m21.stream.Measure(number=j+1)
159                     )
160         ts = m21.meter.TimeSignature('4/4')
161         measures[i][0].insert(0, ts)
162         for k in range(4):
163             # create a string with the note name
164             # and the octave number
165             octave = str(invndx+int(BASENOTE/12))
166             notestr = notes[notelist[invndx][(j*4)+k]]+\
```

```
167                              octave
168              thisdur = durlist[invndx][(j*4)+k]
169              notedata = get_notedata(thisdur)
170              for ndx, noteinfo in enumerate(notedata):
171                  note=m21.note.Note(notestr,
172                                     type=noteinfo[0],
173                                     dots=noteinfo[1])
174                  if ndx == 0 and len(notedata) > 1:
175                      note.tie = m21.tie.Tie("start")
176                  elif ndx == (len(notedata)-1) and \
177                          len(notedata) > 1:
178                      note.tie = m21.tie.Tie("stop")
179                  measures[i][j].append(note)
180          measures[i][len(measures[i])-1].rightBarline='final'
181          parts[i].append(measures[i])
182      return parts
183
184
185  if __name__ == "__main__":
186      # create a random twelve tone line
187      serie = [i for i in range(12)]
188      random.shuffle(serie)
189      # create a string with twelve lines
190      # with unique twelve tone lines
191      # separated with \n
192      series = m21.serial.rowToMatrix(serie)
193      # separate the string above to twelve strings
194      twelvetone_list = series.split('\n')
195      # get a list with twelve lists with integers
196      # from the strings above
197      matrix = []
198      counter = 0
199      for line in twelvetone_list:
200          matrix.append([])
201          splitline = line.split(' ')
202          for string in splitline:
203              if len(string) > 0:
204                  matrix[counter].append(int(string))
205          counter += 1
206      # flatten the 2D list to one 1D list
207      matrix_flat = []
208      for serie in matrix:
209          for note in serie:
210              matrix_flat.append(note)
211
212      listlen = int(len(matrix_flat)/NUMOSC)
213      notelist = [matrix_flat[i*listlen:(i+1)*listlen]
214                  for i in range(NUMOSC)]
215
216      durlist = []
217      masklist = []
218      for i in range(NUMOSC):
```

```
219              masklist.append([])
220              durlist.append([])
221              for j in range(int(len(notelist[i])/METER)):
222                  masklocal = rand_durs(METER, MINDUR)
223                  for dur in masklocal:
224                      masklist[i].append(1)
225                      for k in range(dur-1):
226                          masklist[i].append(0)
227                  durlist[i].append(dur)
228
229          # set the maximum count based on the mask list
230          pyocounter.setMax(len(masklist[0]))
231          select.setValue(len(masklist[0])-1)
232
233          schoenberg = [Schoenberg(i,notelist[i],masklist[i],
234                          durlist[i]) for i in range(NUMOSC)]
235          mix = pyo.Mix(schoenberg, voices=2).out()
236
237          score = m21.stream.Score(id="Schoenberg Style Piece")
238          parts = store_parts(notelist, durlist)
239
240          for i in range(NUMOSC):
241              score.insert(0, parts[i])
242
243          s.start()
244          ca = pyo.CallAfter(metro.play, time=5)
245          tf = pyo.TrigFunc(select, metro.stop)
246
247          score.show()
```

9.3   Playing Bach with Oscillators

The next project we will realise is a script that draws a Bach score from musc21's corpus and plays it back using PyoObjects. In this project we will work in a reverse order, as we need to draw music21 information and translate it into data for PyoObjects. We will realise only one version of this project with the score display included.

Ingredients:

- Python3
- The Pyo module
- The music21 module
- Finale, Sibelius, or MuseScore
- A text editor (preferably an IDE with Python support)
- A terminal window in case your editor does not launch Python scripts

Process:

The code of this project is less than the code of the previous one, so it is displayed in one chunk, shown in Script 9.13. In this project, we apply a different approach to the sequencing mechanism. Since we use an `Adsr()` object to control the amplitude of our audio source, we have to deal with Python's limited timing accuracy, as setting this class's duration is done with a float only, and not with a PyoObject. This should normally work fine for the tempi of this project, but if we get a few voices that all play sixteenths simultaneously, we will likely get clicks as Python suffers from timing accuracy in fast paces. To remedy this, instead of setting a duration for the envelope, we set it to 0, which means that the envelope will stay active in its sustain stage, until its `stop()` method is called.

For this approach to work, we have to process our masking list so that it doubles its size, and before each attack, we place a value that will call the `stop()` method. Since we will double the masking list size, we must also double the speed of the metronome. This way, we will be able to play even the shortest notes of the score, and still be able to stop the envelope before the next note is triggered. We do this because we can set the `play()` and `stop()` methods to be called by `TrigFunc()` objects. This way, the callbacks will happen in the C code of the class, therefore it is very efficient timewise.

Script 9.13 The Bach oscillator playback script.

```
1  import music21 as m21
2  import pyo
3  import random
4
5  s = pyo.Server().boot()
6
7  metro = pyo.Metro()
8  pyocounter = pyo.Counter(metro)
9  select = pyo.Select(pyocounter)
10
11 class BachOsc(pyo.PyoObject):
12     def __init__(self,num_parts=1,parsed_notes=None,
13                     masklist=None,mul=1,add=0):
14         pyo.PyoObject.__init__(self, mul, add)
15         # convert arguments to lists
16         mul,add,lmax = pyo.convertArgsToLists(mul,add)
17         self._mask = pyo.Iter(metro, choice=masklist)
18         self._select_on = pyo.Select(self._mask, value=1)
19         self._select_off = pyo.Select(self._mask, value=2)
20         self._adsr = pyo.Adsr(mul=1/num_parts)
21         self._tf_on = pyo.TrigFunc(self._select_on,
22                                     self._adsr.play)
23         self._tf_off = pyo.TrigFunc(self._select_off,
24                                     self._adsr.stop)
25         self._freqs = pyo.Iter(self._select_on,
26                                 choice=parsed_notes)
27         self._out = pyo.FM(carrier=self._freqs,ratio=.5012,
28                             index=1,mul=self._adsr)
29         self._base_objs = self._out.getBaseObjects()
30
31
32 def parse_score(num_parts, part_stream):
33     parsed_notes = [[]for i in range(num_parts)]
34     parsed_durs = [[]for i in range(num_parts)]
```

```
35      # set a whole note as a starting point
36      metro_time = 4
37      for i in range(num_parts):
38          for note in part_stream[i].flat.notesAndRests:
39              if type(note) == m21.note.Note:
40                  parsed_notes[i].append(note.pitch.frequency)
41              elif type(note) == m21.note.Rest:
42                  parsed_notes[i].append(-1)
43              if note.duration.quarterLength < metro_time:
44                  metro_time = note.duration.quarterLength
45              parsed_durs[i].append(
46                              note.duration.quarterLength
47                          )
48      return parsed_notes, parsed_durs, metro_time
49
50
51 def get_num_steps(parsed_durs,metro_time):
52      # all parts have the same number of steps
53      # so we query the first one only
54      num_steps = 0
55      for i in parsed_durs[0]:
56          num_steps += int(i/metro_time)
57      return num_steps
58
59
60 def add_stop_vals(l):
61      # store the indexes with a 1
62      ones_ndx = []
63      if l[0] == 1:
64          ones_ndx.append(0)
65      for i in range(len(l)):
66          for j in range(i, len(l)):
67              if l[i] == 1 and j > i and l[j] == 1:
68                  ones_ndx.append(j)
69                  break
70      # then double the size of the list
71      for i in range(len(l)):
72          l.append(0)
73      # and double the values of the stored indexes
74      for i in range(len(ones_ndx)):
75          ones_ndx[i] *= 2
76      # place the ones in the new indexes
77      for i in range(len(l)):
78          l[i] = 0
79          if i in ones_ndx:
80              l[i] = 1
81      # and a 2 (the stop value) before each 1
82      for i in ones_ndx:
83          if i > 0:
84              l[i-1] = 2
85      return l
86
```

```
87
88 def get_mask(num_parts, num_steps, metro_time):
89     part_events = [[] for i in range(num_parts)]
90     masklist = [[] for i in range(num_parts)]
91     for i in range(num_parts):
92         event = 0
93         part_events[i].append(0)
94         for j in parsed_durs[i]:
95             event += j/metro_time
96             part_events[i].append(int(event))
97     for i in range(num_parts):
98         for j in range(num_steps):
99             if j in part_events[i]:
100                 masklist[i].append(1)
101             else:
102                 masklist[i].append(0)
103     # iterate through the list and zero any entry with a rest
104     ndx = 0
105     for i in range(len(masklist)):
106         if masklist[i] == 1:
107             if parsed_notes[ndx] < 0:
108                 masklist[i] = 0
109             ndx += 1
110     # double masklist to include a 2
111     # for stopping the envelope
112     # we need to double so that even
113     # the minimum duration can work
114     for part in range(num_parts):
115         masklist[part] = add_stop_vals(masklist[part])
116         # add a 2 at the end to stop when done
117         masklist[part][len(masklist[part])-1] = 2
118     return masklist
119
120
121 if __name__ == "__main__":
122     allBach = m21.corpus.search('bach')
123     x = allBach[random.randrange(len(allBach))]
124
125     bach = x.parse()
126     part_stream = bach.parts.stream()
127     parts = []
128     num_parts = len(part_stream)
129     # get the tempo from the score
130     tempo = bach.metronomeMarkBoundaries()[0][2].number
131     # parse the notes from the score
132     parsed_notes,parsed_durs,metro_time=parse_score(
133                                         num_parts,
134                                         part_stream
135                                         )
136     # get the total number of steps of the sequencer
137     num_steps = get_num_steps(parsed_durs,metro_time)
138
```

```
139        masklist = get_mask(num_parts,num_steps,metro_time)
140        # halve the time to stay in tempo with the new mask list
141        metro_time /= 2
142        metro.setTime(pyo.beatToDur(1/(1/metro_time),tempo))
143        pyocounter.setMax(len(masklist[0])+1)
144        select.setValue(len(masklist[0]))
145        tf = pyo.TrigFunc(select, metro.stop)
146
147        bach_osc = [BachOsc(num_parts,parsed_notes[i],
148                    masklist[i]) for i in range(num_parts)]
149        mix = pyo.Mix(bach_osc, voices=2).out()
150
151        s.start()
152
153        ca = pyo.CallAfter(metro.play, time=5)
154        bach.show()
```

In the `parse_score()` function, we retrieve the musical data from the music21 score. We iterate over the `stream()` object of the passed argument in line 37, and check if each element is a note or a rest. If it is a note, we store the frequency, in line 40, otherwise we store a -1 for a rest, which is filtered later on. We also store the duration of every note or rest, based on the length of a quarter note. If a note is indeed a quarter note, its duration will be 1. If it is an eighth note, its duration will be 0.5, and so on.

The `get_mask()` function sets the values for the masking list that is used for the sequencing, and the `add_stop_vals()` function sets the stop values that were mentioned above. The discussion on these functions is intentionally skipped so the reader can work out how these work, for themselves. The comments in the code of the latter should help understanding. Wherever you find difficulty understanding mathematical operations, suppose that the variables used have the simplest values possible. For example, in the `get_mask()` function, suppose that the `metro_time` variable is 1, as if the shortest duration of a piece is the quarter note.

The rest of the code should be easy to understand by now. Something we haven't seen so far, is line 130, where we retrieve the tempo of the piece. The `metronomeMarkBoundaries()` method returns a list with information about the metronome, where the third element is the BPM. We can retrieve this information the way it is done in line 130. If this information is absent, then a default value of 120 is passed for the BPM.

In line 147 we create a list of `BachOsc()` objects, instead of passing the lists and creating a multi-stream single object, in the style of Pyo. This is because if we create one object and pass the lists of lists without indexing, the ADSR envelopes will all fire together. This means that if a voice has an eighth note and another voice has a sixteenth note, they will both fire at the tempo of the sixteenth, with the former voice playing the same note twice. By creating a list of objects, we remedy this issue and all voices are independent of each other.

The `pyocounter` that keeps track of the sequencer steps and stops the metronome when the piece is over, needs to count one step too many, otherwise the oscillators will hang at the end and will not stop. By adding one more step to the count, the `stop()` method of the ADSR envelopes is called and the oscillators stop sounding when the piece is over.

Note that this script is generic, as it randomly chooses a piece from the Bach corpus of music21. There are some pieces that might raise errors, depending on the type of information included in the chosen piece. If you get an error, run the code again to choose another piece that will be more likely to work. Most pieces work with this generic code.

9.4   Conclusion

We have seen how we can draw information from the music21 module that is useful in a compositional context. We have also seen how convenient Object Oriented Programming (OOP) can be, by abstracting code in classes and using them one inside another. The two projects realised in this chapter should mainly serve as inspiration for further endeavours in the field of algorithmic composition. If you are a more traditional composer, you might feel at home when working with the music21 module, to create computer music. If you are more of an electronic music composer, you still might find interesting resources in this module.

Music21 is a very flexible toolkit for musicology, but can also serve as a toolkit for composition, as it provides a vast array of tools and a very big corpus of compositions one can use. We have only touched the surface of this module, but you are encouraged to search deeper. The documentation of music21[3] is well written and has great detail. Be it only to create sheet music out of algorithmic compositions realised with other means, use its corpus to train Recurrent Neural Networks that will then generate music of merged styles, or other approaches, this module has the potential to become one of the main tools of a composer.

9.5   Exercises

9.5.1   Exercise 1

Change the sequencing mechanism of the first project to the one of the second project. Instead of setting a duration for the ADSR envelope, change the masking lists so that the `play()` and `stop()` methods of the `Adsr()` class are called separately.

9.5.2   Exercise 2

Change the sound source of the second project with a class you define yourself, similar to the first project.

Tip: Make sure you call the `getBaseObjects()` method for the PyoObject you want your class to treat as the output.

9.5.3   Exercise 3

Modify the Bach playback script to treat ties.

Tip: A tie is a subclass of the `note.Note()` class. In the `parse_score()` function, where we iterate over the `notesAndRests` of the part streams, check if a note has a tie by querying if `note.tie is not None`. If this is the case, you can get the state of the tie with `note.tie.type` which returns one of the following strings: "`start`", "`continue`", or "`stop`". You have to take care to modify the masking list so that tied notes are not triggered, but also not stopped before the end of the last tied note.

Notes

1 https://stackoverflow.com/questions/3589214/generate-random-numbers-summing-to-a-predefined-value
2 http://web.mit.edu/music21/doc/usersGuide/usersGuide_08_installingMusicXML.html
3 https://web.mit.edu/music21/doc/index.html

Bibliography

Cuthbert, M. and Ariza, C. (2010) 'music21: A Toolkit for Computer-Aided Musicology and Symbolic Music Data.', in *Proceedings of the 11th International Society for Music Information Retrieval Conference, ISMIR 2010*. Utrecht, The Netherlands, pp. 637–642.

Haimo, E. (1993) *Schoenberg's Serial Odyssey: The Evolution of his Twelve-Tone Method, 1914–1928*. Oxford, UK: Oxford University Press.

10 The Events Framework

Pyo includes a framework for sequencing musical events, called the Events Framework. This framework consists of a core class that computes parameters specific to this framework, and a number of classes that add various functionalities. Even though Pyo provides other means for creating sequences, the Events Framework provides a high level of abstraction where users can create complex structures with minimal coding.

What You Will Learn

After you have read this chapter you will:

- Know the basics of the Events Framework
- Know how to implement complex musical structures with minimal coding
- Be able to approach previous projects with the Events Framework

---

10.1 The Events() and EventSeq() Classes

The Events Framework is built around the `Events()` class. This class takes various kwargs and computes the passed information to create sequences. In the simplest form, this class is combined with the `EventSeq()` class, to iterate linearly over a given sequence.

Ingredients:

- Python3
- The Pyo module
- An interactive Python shell

Process:

Let us create a simple sequence of notes with specified durations. Shell 10.1 shows this simple sequence. The fourth line creates an arpeggio of a C major 7th chord, with durations of one quarter note for the root, one eighth note for E, and two sixteenth notes for G and B.

DOI: 10.4324/9781003386964-10

Shell 10.1 A basic sequence with the Events Framework.

```
>>> from pyo import *
>>> s = Server().boot()
>>> s.start()
>>> e = Events(
...      midinote=EventSeq([60,64,67,71]),
...      beat=EventSeq([1,1/2,1/4,1/4])
... )
>>> e.play() # let it play some time, note that sound will be loud
>>> e.stop()
```

As the comment in this shell session states, the sound will be at full volume, so take care to reduce the volume of your speakers. The Events() class takes many kwargs to control various parameters. To control the amplitude of its output, we can pass one of the following three kwargs: amp, db, midivel. The first kwarg takes values from 0 to 1. The second one expresses the amplitude in the dB range, where 0 is the maximum amplitude, and 10dB is roughly considered to double or halve the amplitude. The third kwarg, as you can imagine, expresses loudness in MIDI velocity values, between 0 and 127.

As with the amplitude, the pitch can also be expressed in three different ways, with the following kwargs: freq, midinote, degree. The first kwarg expresses pitch in raw frequency values. The second one – used in Shell 10.1 – expresses pitch in MIDI note numbers, where 60 is the middle C, and 69 is 440Hz. The third kwarg expresses pitch in float values, where the integral part is the octave, and the decimals are the semitones. For example, the value 5.07 is seven semitones in the fifth octave, equal to MIDI 67. The range of the decimals is between 00 and 11.

The durations of notes are expressed in fractions of a quarter note. A 1 is a quarter note, a 2 is a half note, and a 1/2 is an eighth note. A useful kwarg that connects to the durations is durmul. This kwarg sets a coefficient that is multiplied by the beat attribute, letting the user control the length of a note, without affecting the overall timing. For example, a value of 0.5 will create short, staccato-like notes, as the durations set by the beat kwarg will be halved, whereas a value of 1.5 will create an overlapping of notes, as the durations will last longer than their beat.

In Shell 10.1, we use the EventSeq() class in both kwargs of the Events() object we created. Instead of this class, we can pass a simple list or a single value. In this case, whether a list or a single value, the respective kwarg will be converted internally to an EventSeq() object.

10.1.1   Setting the Number of Repetitions

The EventSeq() class includes an occurrences attribute that sets the number of repetitions of a list. Shell 10.1 can change to Shell 10.2 to repeat this sequence twice and then stop.

Shell 10.2 Setting a number of occurrences to EventSeq().

```
>>> from pyo import *
>>> s = Server().boot()
>>> s.start()
>>> e = Events(
...      degree=EventSeq([5.00,5.04,5.07,5.11]),
```

```
...        beat=EventSeq([1,1/2,1/4,1/4],occurrences=2),
...        amp=0.2,durmul=0.5
... ).play()
```

In this shell we also use the `degree` kwarg to set the notes, `amp` to lower the overall amplitude, and `durmul` to create a staccato feel. The notes in Shell 10.2 are the same with Shell 10.1. To stop a sequence after a certain number of repetitions, we have to use the `occurrences` kwarg of `EventSeq()` for the `beat` kwarg of `Events()`, because this attribute triggers an internal envelope. If the envelope stops being triggered, the sound will also stop. If another sequence stops being triggered, like the frequencies, the sequence will keep on going, and the `Event-Seq()` that has stopped will just repeat its last value over and over.

10.1.2   Setting an EventSeq() Object for any Events() Kwarg

It is easy to imagine that any `Events()` parameter can take an `EventSeq()` instead of a single value. We can work on Shell 10.2 and set a sequence for both the `amp` and the `durmul` kwargs. Shell 10.3 does this.

Shell 10.3 Using EventSeq() objects for all kwargs.
```
>>> from pyo import *
>>> s = Server().boot()
>>> s.start()
>>> e = Events(
...        degree=EventSeq([5.00,5.04,5.07,5.11]),
...        beat=EventSeq([1,1/2,1/4,1/4],occurrences=2),
...        amp=EventSeq([0.05,0.1,0.15,0.08]),
...        durmul=EventSeq([0.5,0.8,1.2,0.5])
... ).play()
```

10.1.3   Controlling the Envelope of an Events() Object

It is possible to control the amplitude envelope of the `Events()` class either by setting values for all or part of the stages of an ADSR envelope, or by passing a PyoTableObject to the `envelope` kwarg. If we pass ADSR stages as arguments, we can pass all four, or skip the decay stage, in which case we will get an ASR (Attack-Sustain-Release) envelope. If we need a more specific envelope, we can create it with any PyoTableObject, like `CosTable()`, `ExpTable()`, or any other table class. Shells 10.4 and 10.5 show code for these two versions.

Shell 10.4 Controlling the amplitude envelope with ADSR values.
```
>>> from pyo import *
>>> s = Server().boot()
>>> e = Events(
...        degree=EventSeq([5.00,5.04,5.07,5.11]),
...        beat=EventSeq([1,1/2,1/4,1/4],occurrences=2),
...        amp=0.2,attack=0.001,decay=0.05,sustain=0.6,release=0.02
... ).play()
```

Shell 10.5 Controlling the amplitude with a PyoTableObject.
```
>>> from pyo import *
>>> s = Server().boot()
```

```
>>> env=CosTable([(0,0.0),(256,1.0),(1024,0.6),
...                 (7680,0.6),(8192,0.0)])
>>> e = Events(
...     degree=EventSeq([5.00,5.04,5.07,5.11]),
...     beat=EventSeq([1,1/2,1/4,1/4],occurrences=2),
...     amp=0.2,envelope=env
... ).play()
```

When setting explicit values for the ADSR stages, we have to be careful to provide values that sum up to the total duration of the shortest note in the sequence, otherwise it is possible to either get clicks, or overlapping tones. With a PyoTableObject this is not the case, as the Events() class takes care of the duration of the envelope.

10.1.4 Other Sequencing Classes

The Events Framework includes more sequencing classes. These are EventSlide(), EventIndex(), EventMarkov(), EventChoice(), EventDrunk(), Event-Noise(), and EventConditional(). The EventChoice(), EventDrunk(), and EventNoise() classes provide different levels of randomness in the sequence. Event-Markov() applies a Markov chain in the sequence, based on its input list. EventIndex() plays values based on a given index. The index can be another EventGenerator, or a PyoObject, so this class can behave from very linear to completely random. EventSlide() can produce repetitive patterns, as it plays a segment of a sequence, then goes back a number of steps, speci-fied by the user, and repeats this pattern in a loop.

EventConditional() is not really a sequencing class, but combined with other Event-Generator sequencers, it can modify a sequence. It performs a conditional test and outputs its iftrue value if the condition is True, and its iffalse if the condition is False. True and False is anything Python considers to be True or False.

Documentation Pages:

Events(): https://belangeo.github.io/pyo/api/classes/events.html#events
EventSeq(): https://belangeo.github.io/pyo/api/classes/events.html#eventseq
EventSlide(): https://belangeo.github.io/pyo/api/classes/events.html#eventslide
EventIndex(): https://belangeo.github.io/pyo/api/classes/events.html#eventindex
EventMarkov(): https://belangeo.github.io/pyo/api/classes/events.html#eventmarkov
EventChoice(): https://belangeo.github.io/pyo/api/classes/events.html#eventchoice
EventDrunk(): https://belangeo.github.io/pyo/api/classes/events.html#eventdrunk
EventNoise(): https://belangeo.github.io/pyo/api/classes/events.html#eventnoise
EventConditional(): https://belangeo.github.io/pyo/api/classes/events.html#eventconditional
CosTable(): https://belangeo.github.io/pyo/api/classes/tables.html#costable

10.2   Using Our Own Audio Classes

As you have noticed, the sound produced by the Events Framework is always the same. This is because this framework uses a stereo RCOsc() class by default. It is possible and quite easy to replace this with a custom class, so we can use this framework with any sound we like. We saw how to write our own classes in Chapter 8, and in Chapter 9 we practiced on this concept in more realistic scenarios. Let's now see how we can apply our own classes in the Events Framework.

Ingredients:

- Python3
- The Pyo module
- The wxPython module, if you want to use Pyo's GUI
- A text editor (preferably an IDE with Python support)
- A terminal window in case your editor does not launch Python scripts

Process:

To be able to write our own classes inside the Events Framework, we must inherit from the `EventInstrument()` class. By doing so, there are certain attributes that our class will inherit, including the frequency, duration, and amplitude. The `EventInstrument()` parent class also takes care to clear its resources when done playing. Script 10.1 creates the triangle wave oscillator with breakpoint control that we created in Chapter 8, modified so it can be used in the Events Framework.

Script 10.1 A triangle wave oscillator as the audio class of the Events Framework.

```
1 from pyo import *
2 s = Server().boot()
3
4 class EventTri(EventInstrument):
5     def __init__(self, **kwargs):
6         EventInstrument.__init__(self, **kwargs)
7         self._brkpnt = Sig(self.brkpnt)
8         self._invbrk = 1.0 - self._brkpnt
9         self._phasor = Phasor(freq=self.freq)
10        self._rising = (self._phasor/self._brkpnt) * \
11             (self._phasor < self._brkpnt)
12        self._falling = (((self._phasor-self._brkpnt)/\
13             self._invbrk)*(-1)+1)*(self._phasor>=self._brkpnt)
14        self._osc=Sig((self._rising+self._falling),
15                      mul=2, add=-1)
16        self.output = Sig(self._osc, mul=self.env).out()
17
18
19 event = Events(
20     instr=EventTri,
21     degree=EventSeq([5.00, 5.04, 5.07, 6.00]),
22     brkpnt=EventSeq([0.2, 0.4, 0.6, 0.8]),
23     beat= 1/2
24 ).play()
25
26 s.gui(locals())
```

Lines 4 to 6 are necessary for our class to work within the Events Framework. The only parameter from these lines that can change is the name of the class. The body of the class should look familiar. Note though that we do not define any `self.freq` or `self.env` attributes, even though we use them in lines 9 and 16 respectively. These attributes are derived from the parent class, and are either passed as kwargs – the `degree` kwarg for the frequency – or created internally by the `Events()` class – in this case, the default envelope.

On the other hand, the `brkpnt` kwarg passed in the `Events()` class is not a kwarg this class expects. This is a kwarg specific to our own class, used in line 7. The Events Framework enables us to pass arbitrary kwargs so we can modify any parameter of our classes as part of this framework. For this specific kwarg, we pass an `EventSeq()` object. This is possible because our class inherits from the `EventInstrument()` class, making it part of this ecosystem.

To be able to use our class in the Events Framework, we need to pass the name of the class to the `instr` kwarg. When defining classes to be used within the Events Framework, we don't need to call the `getBaseObjects()` method for the object we want to use as the audio output. We just call its `out()` method, like we would do when creating PyoObjects outside of a class. This is done in line 16.

Documentation Page:

`EventInstrument()`: https://belangeo.github.io/pyo/api/classes/events.html#eventinstrument

10.3   Calling Functions

It is possible to call functions within the Events Framework. These functions can either be built in Python functions – like the various functions of the random module – or functions that we define ourselves.

Ingredients:

- Python3
- The Pyo module
- The wxPython module, if you want to use Pyo's GUI
- A text editor (preferably an IDE with Python support)
- A terminal window in case your editor does not launch Python scripts

Process:

The `EventCall()` class enables us to call any function from within the Events Framework. We can pass any number of positional and keyword arguments to our function, and return a value that will be used by the kwarg of the `Events()` class for which we assign an `Event-Call()` object. Script 10.2 calls the `clipfreq()` function with random input provided by the `EventNoise()` class. The result of this script is to play a MIDI note 60 – middle C – if the random input to the function is less than 60, otherwise play the random input.

Script 10.2 Calling a function within the Events Framework.
```
1 from pyo import *
2 s = Server().boot()
```

```
3
4 def clipfreq(val):
5      if val > 60:
6           return val
7      else:
8          ·return 60
9
10 event = Events(
11      midinote=EventCall(
12                  clipfreq,EventNoise().rescale(-1,1,48,72,1)
13              ),
14      beat= 1/4,
15      db=-12,
16      attack=0.001,
17      decay=0.05,
18      sustain=0.5,
19      release=0.005,
20 ).play()
21
22 s.gui(locals())
```

The first argument to the `EventCall()` class is the function to be called, after that we can provide any positional arguments the functions expects, and after that we can include any kwargs we might want to use. The single positional argument the `clipfreq()` function expects, is provided by the `EventNoise()` class, after it is rescaled from its default range from -1 to 1, to a range from 48 to 72, with an exponent of 1, resulting in a linear output.

Documentation Page:

`EventCall()`: https://belangeo.github.io/pyo/api/classes/events.html#eventcall

10.4   Nesting Events Generators

As we saw in the previous example, some EventGenerators can take other EventGenerators as arguments. For example, we can pass a list of `EventSeq()` objects to the `values` argument of another `EventSeq()` object.

Ingredients:

- Python3
- The Pyo module
- The wxPython module, if you want to use Pyo's GUI
- A text editor (preferably an IDE with Python support)
- A terminal window in case your editor does not launch Python scripts

Process:

If we set a definite number of `occurrences` in the inner `EventSeq()`'s, these will repeat their values each in turn, until their number of `occurrences` is met. Script 10.3 creates the following MIDI note sequence: [[60, 62, 63], [60, 62, 63]], [[62, 67], [62, 67], [62, 67]], [[65, 64, 62, 66]], [[67, 62], [67, 62]]. The outter square brackets are used to distinguish the sequences within the nested `EventSeq()`'s, and the inner brackets to distinguish each occurrence separately. There is no limit to the number of EventGenerators that can be nested.

Script 10.3 Nesting EventGenerators.

```
1 from pyo import *
2 s = Server().boot()
3
4 event = Events(
5     midinote=EventSeq(
6         [
7             EventSeq([60, 62, 63], occurrences=2),
8             EventSeq([62, 67], occurrences=3),
9             EventSeq([65, 64, 62, 66], occurrences=1),
10            EventSeq([67, 62], occurrences=2),
11        ],
12        occurrences=inf,
13    ),
14    beat= 1/4
15 ).play()
16
17 s.gui(locals())
```

10.5   Sharing Values between Generators

Sharing values within an EventGenerator or even between different EventGenerators is possible with the Events Framework.

Ingredients:

- Python3
- The Pyo module
- The wxPython module, if you want to use Pyo's GUI
- A text editor (preferably an IDE with Python support)
- A terminal window in case your editor does not launch Python scripts

Process:

We can share the values of a kwarg within an `Events()` object the way it is done in line 13 in Script 10.4. The `EventKey()` class provides access to a kwarg by passing the name of the kwarg as a string to its first positional argument. In Script 10.4 we assign a random note to the `midinote` kwarg, and then we use this random note as input to the `setamp()` function, the returned value of which we pass to the `db` kwarg. The result is that notes from MIDI value 60 and above will be played louder than notes below this threshold.

Script 10.4 Sharing values between EventGenerators.

```
1 from pyo import *
2 s = Server().boot()
```

```
3
4 def setamp(val):
5      if val > 60:
6           return -12
7      else:
8           return -24
9
10 event1 = Events(
11      midinote=EventNoise().rescale(-1,1,48,72,1),
12      beat= 1/4,
13      db=EventCall(setamp,EventKey("midinote"))
14 ).play()
15
16 event2 = Events(
17      midinote=EventKey(
18                "midinote",master=event1
19              ).rescale(48,72,72,48,1),
20      beat=1/4,
21      db=EventCall(setamp,EventKey("midinote"))
22 ).play()
23
24 s.gui(locals())
```

In the second Events() object of this script, we access the midinote kwarg of the first Events() object, by passing the name of the later to the master kwarg of the Event-Key(). We then rescale the output so the two Events() will play mirrored pitches. Then, in line 21, we access the midinote kwarg of the event2 object inside itself – as we don't provide a value to the master kwarg of EventKey() – and apply the same process for setting the amplitude in dB that we did with event1. Make sure you understand which midinote kwarg of which Events() class is passed to each EventKey() object. Only in lines 17 to 19 do we access the kwarg of another Events() class. All other EventKey() objects access kwargs from the Events() class inside of which they are created.

Documentation Page:

EventKey(): https://belangeo.github.io/pyo/api/classes/events.html#eventkey

10.6   Using Other PyoObjects as Input to Generators

So far we have been providing lists or EventGenerators to all kwargs of all EventGenerators. We can pass any PyoObject to a kwarg though.

Ingredients:

- Python3
- The Pyo module
- The wxPython module, if you want to use Pyo's GUI
- A text editor (preferably an IDE with Python support)
- A terminal window in case your editor does not launch Python scripts

Process:

We can revisit the breakpoint triangle oscillator and use a sine wave to control its break-point, instead of an `EventSeq()`. In Script 10.5, we create a sine wave and pass that to the `brkpnt` kwarg of our event. This PyoObject could be passed to any kwarg, including the ones that are part of the `Events()` class, and not only to additional kwargs passed to the instrument class.

Script 10.5 Passing a PyoObject to an Events() kwarg.

```
1 from pyo import *
2 s = Server().boot()
3
4 class EventTri(EventInstrument):
5      def __init__(self, **kwargs):
6          EventInstrument.__init__(self, **kwargs)
7          self._brkpnt = Sig(self.brkpnt)
8          self._invbrk = 1.0 - self._brkpnt
9          self._phasor = Phasor(freq=self.freq)
10         self._rising = (self._phasor/self._brkpnt) * \
11             (self._phasor < self._brkpnt)
12         self._falling = (((self._phasor-self._brkpnt)/\
13            self._invbrk)*(-1)+1)*(self._phasor>=self._brkpnt)
14         self._osc = Sig((self._rising + self._falling),
15                    mul=2, add=-1)
16         self.output = Sig(self._osc, mul=self.env).out()
17
18 sine = Sine(freq=.4,mul=.4,add=.5)
19
20 event = Events(
21     instr=EventTri,
22     degree=EventSeq([5.00, 5.04, 5.07, 6.00]),
23     brkpnt=sine,
24     beat=1/2
25 ).play()
26
27 s.gui(locals())
```

10.7   Accessing the Audio Output of Events() in Other PyoObjects

As it is possible to provide the stream of any PyoObject as input to an `Events()` object, it is also possible to access the signal output of an `Events()` object in any other PyoObject. This process is slightly more involved than passing PyoObject streams to EventGenerators, but it is still rather simple.

Ingredients:

- Python3
- The Pyo module
- The wxPython module, if you want to use Pyo's GUI

- A text editor (preferably an IDE with Python support)
- A terminal window in case your editor does not launch Python scripts

Process:

We can develop the previous script, so we can access its audio stream to display it in Pyo's Scope (). The code is shown in Script 10.6. In this version, we are creating a stereo output by calling the mix () method in line 16. To get access to the audio output of an Events () object, we need to specify two kwargs, signal and outs. The first one takes a string with the name of the object we want to output. This is the self.output object in line 16. If we don't specify a class to be used as the instrument and we use the default instrument of the Events Framework, then the argument that should be passed to the signal kwarg is "sig".

Script 10.6 Accessing the audio output of an Events() object.

```
1 from pyo import *
2 s = Server().boot()
3
4 class EventTri(EventInstrument):
5     def __init__(self, **kwargs):
6         EventInstrument.__init__(self, **kwargs)
7         self._brkpnt = Sig(self.brkpnt)
8         self._invbrk = 1.0 - self._brkpnt
9         self._phasor = Phasor(freq=self.freq)
10        self._rising = (self._phasor/self._brkpnt) * \
11              (self._phasor < self._brkpnt)
12        self._falling = (((self._phasor-self._brkpnt)/\
13            self._invbrk)*(-1)+1)*(self._phasor>=self._brkpnt)
14        self._osc = Sig((self._rising + self._falling),
15                        mul=2, add=-1)
16        self.output = Sig(self._osc,mul=self.env).mix(2).out()
17
18 sine = Sine(freq=.4,mul=.4,add=.5)
19
20 event = Events(
21     instr=EventTri,
22     degree=EventSeq([5.00, 5.04, 5.07, 6.00]),
23     brkpnt=sine,
24     signal="output",
25     outs=2,
26     beat=1/2
27 ).play()
28
29 sc = Scope(event.sig())
30
31 s.gui(locals())
```

The outs kwarg takes the number of outputs expected by the Events () object. In Script 10.6 we set the output of the EventTri () class to stereo, so the value passed to the outs kwarg is 2. The default instrument of the Events Framework also produces a stereo output, so if that is used, again the same value should be passed here. Once we set these two kwargs, we can access the output of the Events() object by calling its sig() method, the way it is done in line 29.

10.8 Revisiting Bach

Now that we have covered the core structure of the Events Framework, it is a good exercise to try and replicate the project from the previous chapter that plays back Bach music with oscillators, this time using this framework.

Ingredients:

- Python3
- The Pyo module
- The music21 module
- Finale, Sibelius, or MuseScore
- A text editor (preferably an IDE with Python support)
- A terminal window in case your editor does not launch Python scripts

Process:

The code for this version is much shorter than the one from Chapter 9, as the Events Framework takes care of a lot of calculations that are necessary for sequencing. The code is shown in Script 10.7.

Script 10.7 Playing back Bach music with the Events Framework.

```
1 import pyo
2 import music21 as m21
3 import random
4
5 s = pyo.Server().boot()
6
7 class BachOsc(pyo.EventInstrument):
8     def __init__(self, **kwargs):
9         pyo.EventInstrument.__init__(self, **kwargs)
10        self.output=pyo.FM(carrier=self.freq,
11                           ratio=.5012,index=1,
12                           mul=self.env*self.mul).mix(2).out()
13
14
15 def parse_score(num_parts, part_stream):
16     parsed_notes = [[] for i in range(num_parts)]
17     parsed_durs = [[] for i in range(num_parts)]
18     for i in range(num_parts):
19         for note in part_stream[i].flat.notesAndRests:
20             if type(note) == m21.note.Note:
21                 parsed_notes[i].append(note.pitch.frequency)
22             elif type(note) == m21.note.Rest:
23                 parsed_notes[i].append(-1)
24             parsed_durs[i].append(note.duration.quarterLength)
25     return parsed_notes, parsed_durs
```

```
26
27
28  if __name__ == "__main__":
29      allBach = m21.corpus.search('bach')
30      x = allBach[random.randrange(len(allBach))]
31
32      bach = x.parse()
33      part_stream = bach.parts.stream()
34      num_parts = len(part_stream)
35      # get the tempo from the score
36      tempo = bach.metronomeMarkBoundaries()[0][2].number
37      # parse the notes from the score
38      parsed_notes,parsed_durs=parse_score(
39                               num_parts,part_stream
40                          )
41
42      events = [pyo.Events(
43          instr=BachOsc,
44          mul=1/num_parts,
45          freq=pyo.EventSeq(parsed_notes[i]),
46          beat=pyo.EventSeq(parsed_durs[i], occurrences=1),
47          bpm=tempo
48      ).play(delay=5) for i in range(num_parts)]
49
50      s.start()
51
52      bach.show()
```

There are only a few things we need to explain in this code. The `EventInstrument` we define is minimal. We provide a stereo output in this version, and we take care to reduce the volume of the output by the inverse of the number of parts of the Bach piece we will parse. The `self.mul` value used in line 12 is passed as an extra kwarg to the `Events()` object, in line 44.

We don't need a masking list, as the `Events()` class takes care of triggering its oscillator, so we omit it in the `parse_score()` function. We have also omitted the `metro_time` value we were returning in the initial version of this project. Conveniently, music21's `note.duration.quarterLength` provides the same values the `beat` kwarg that the `Events()` class is expecting, so we don't need to modify this list at all. Also, the frequencies can be provided as raw Hz values by passing them to the `freq` kwarg, instead of `midinotes` or `degree`, so we pass the list with the values derived by `note.pitch.frequency`. Passing a -1 value for rests works as expected and we get silence when there is a rest in the score.

Once we parse the score, we pass the notes and durations lists to the `events` list of `Events()` objects. We set the instrument we want to use, and the tempo through the `bpm` kwarg. Note that we pass the `delay` kwarg to the `play()` method, as it is easier than trying to delay the initialisation of the sequencers with a `CallAfter()` object.

10.9 Conclusion

We have been introduced to the Events Framework, a very flexible framework for organising musical events within Pyo. Its high level of abstraction enables users to minimise coding to the very essentials, as the classes of this framework take care of most of the tasks necessary to sequence events in a musical context. Despite this high level of abstraction, the Events Framework enables users to specify any desired parameter, including the audio class to be used. The API for writing classes for this framework is simple and intuitive, making class definitions even simpler than the already simple standard Python class definition.

By revisiting the Bach playback project, we saw how this framework can prove to be very useful. The length of the code of this version of this project is less than half the length of the original version, leaving very little of the cumbersome work to the coder. As we saw in the first section of this chapter, the Events Framework includes a variety of sequencing classes, enabling the creation of musical structures that are much more complex than the linear sequencing we have seen here. You are encouraged to read through the documentation of this framework,[1] but also the online tutorials provided in Pyo's website.[2]

10.10   Exercises

10.10.1   Exercise 1

Revisit the twelve-tone project of Chapter 9 and realise it using the Events Framework.

10.10.2   Exercise 2

Treat ties in the Bach playback script, similar to Exercise 3 from Chapter 9.

10.10.3   Exercise 3

Create an envelope with a PyoTableObject and use that for either the Bach playback or the twelve-tone script from Exercise 1. Make sure the envelope goes to silence about one quarter of its length before its end.

Documentation Page:

PyoTableObjects documentation: https://belangeo.github.io/pyo/api/classes/tables.html

10.10.4   Exercise 4

Create some sort of canon in the Bach playback script by using the EventSlide() class for sequencing notes and durations.

10.10.5 *Exercise 5*

Use `EventMarkov()` to create a composition based on a Bach piece (or some other composer you might want to search through the corpus of music21). Check the documentation of `EventMarkov()`.

Notes

1 https://belangeo.github.io/pyo/api/classes/events.html
2 https://belangeo.github.io/pyo/examples/22-events/index.html

11 The MML and Prefix Expression

In this chapter we will explore the two mini-languages that come with Pyo, the Music Macro Language (MML) and the Prefix Expression. These languages provide either an interface for music composition – the MML – or a simple way to create algorithms that need to go down to sample level, since this is not possible with standard Python classes – the Prefix Expression.

What You Will Learn

After you have read this chapter you will:

- Be able to understand the MML and Prefix Expression mini-languages
- Be able to identify when one of these languages can be useful in a project
- Be able to use these languages in your Pyo projects

11.1 What is the MML

The MML is a music description language developed in the early 1980s (Matsushima, 1997, p. 143). It has been widely used in video games and for sequencing music. Slight variations of the language seem to exist, with minor differences – mainly whether letters are written in upper or lower case, or a few other symbols. Overall, MML is a simple, yet musically expressive language, and any variation of it should look familiar to anyone using another variation.

Pyo includes an MML parser, where MML expressions can be written in a multi-line string, and evaluated by the MML() class. Once an MML expression is parsed, this class provides triggers and information on frequency, amplitude, duration, and arbitrary variables, that can be used in any other PyoObject. This way we can write sequences in more musical terms that specifying frequency values and durations in seconds, but by using note names and specifying durations in the Western-music note duration format.

11.2 The MML Language

MML is very simple, and easy to learn. Its approach is closer to the Western-music composer rather than the computer music programmer, using names and concepts that are common between composers of acoustic music. In this section we will learn the entire language, as it is very small.

DOI: 10.4324/9781003386964-11

Ingredients:

- Python3
- The Pyo module
- The wxPython module, if you want to use Pyo's GUI
- A text editor (preferably an IDE with Python support)
- A terminal window in case your editor does not launch Python scripts

Process:

We will start with creating simple notes with their durations and accidentals, and then we will add rests, loops, and other musical elements to our MML strings.

11.2.1   Notes and Durations

The lowercase letters a to g specify notes in the Dutch notation naming system. A numeric value following a note name indicates the duration of the note. Table 11.1 shows the available durations and their numbers in MML.

Table 11.1 Note durations and their numbers in MML.

Note	MML number
32nd note	0
16th note	1
Dotted 16th note	2
8th note	3
Dotted 8th note	4
Quarter note	5
Dotted quarter note	6
Half note	7
Dotted half note	8
Whole note	9

Accidentals are expressed with the plus sign "+" for a sharp, and the minus "-" for a flat. The octave of a note is expressed with the letter "o" followed by the octave number. To increment an octave we must type o+, and to decrement we must type o-. This will increment or decrement by one octave. To increment or decrement for a specific number of octaves, we must type this number right after the plus or minus sign. The specified octave is stored within a sequence until we provide an octave change. The same applies to note durations. Script 11.1 provides an example using all the information we have so far.

Script 11.1 Two simple sequences in MML.

```
1 from pyo import *
2
3 s = Server().boot()
4
5 seq = """
6 A = o5 c3 d e f+ g a b o+ c
```

```
 7  B = o4 b3 o+ c5 o-2 d3 e5 f+3 g
 8  #0 t92 v60 A B
 9  """

10
11  mml = MML(seq, voices=1).play()
12
13  tab = CosTable([(0,0),(64,1),(1024,1),(4096, 0.5),(8191,0)])
14  env = TrigEnv(mml.getVoice(0), table=tab,
15                  dur=mml.getVoice(0, "dur"),
16                  mul=mml.getVoice(0, "amp"))
17  sine = Sine(freq=mml.getVoice(0, "freq"),mul=env).mix(2).out()
18
19  s.gui(locals())
```

The two sequences are defined in lines 6 and 7. The first line plays a G major scale starting from C – so we can consider it to be a C Lydian – in the fifth octave. All notes are eighth notes. The last C is played one octave higher, in the sixth octave. The second sequence starts on the fourth octave and plays a B eighth note. Then it moves one octave up, to the fifth octave, and plays a C quarter note. It then moves two octaves down and plays a D eighth note, an E quarter note, and an F sharp and a G eighth notes. As with octaves, note durations don't need to be repeated if they don't change, so the first line has only one note duration specified.

These two sequences are assigned to the variables A and B, which are called macros. In line 8 we define the voice number with the hash symbol (#). The letter "t" defines the tempo in BPM, and with letter "v" defines the volume in a range between 0 and 100. This voice will play at 92 BPM, 60% of its full amplitude, and it will first play the sequence A and then the sequence B. Once the two sequences are done, it will stop playing.

To get the information of a voice we must first create an MML() object, and pass our MML string to its first argument. We must also set the number of voices, which should match the number of voices of our MML string. In the case of Script 11.1, there is only one voice, so we provide this number to the voices kwarg, which is the default. In lines 13 to 17 we create an envelope in a CosTable(), an envelope trigger, and a sine wave oscillator to play this sequence. To get information on frequency, duration, and amplitude, we must call the get-Voice() method of the MML() class with the voice number as the first argument, and the respective string, as shown in this script. By calling this method with the voice number only, we get the stream of trigger signals. We pass this stream to the first argument of the TrigEnv() object, in line 14, so our envelope gets triggered.

Note that the MML() class keeps durations and octave numbers in its internal states. This means that in Script 11.1, since we first play sequence A and then sequence B, we could omit the duration for the first note in sequence B, since it is the same as the duration of the last note of sequence A. If we inverse the sequences though, the first note in the internal state of MML() will not have a duration assigned to it. To avoid such problems, it is better to always start a sequence by defining the octave number and the duration of the first note.

11.2.2 Rests, Loops, Comments, and Variables

Let's now include some more elements of the MML. Rests are written with the letter "r", and their duration is expressed the same way with the notes. Looping is possible with |: denoting the beginning of a loop, and :| denoting the end. A number following the latter symbol indicates how many times a loop should be repeated. If no number is provided, then the sequence inside the loop symbols will be played twice. Loops can be defined both inside a macro and a voice definition. Script 11.2 plays the first five bars of Györgi Ligeti's *Musica Ricercata III*, twice, as we create loops of macros in lines 14, 17, and 20.

Script 11.2 The first five bars of Györgi Ligeti's *Musica Ricercata III* in MML.

```
 1 from pyo import *
 2
 3 NUM_VOICES = 3
 4
 5 s = Server().boot()
 6
 7 seq = """
 8 ; Title: Musica Ricercata III
 9 ; Composer: Gyorgi Ligeti
10 A0 = o6 |: r3 e-1 c e-3 c e-5 c :| \
11      r3 e-1 c e-3 c e-5 g
12 B0 = r3 e-1 c e-3 c e- c o- g e- \
13      c o- g o+ c e- g o+ c e- g
14 #0 t176 v80 x.08 |: A0 B0 :|
15 A1 = o4 |: c3 r3 r5 r7 :|4
16 B1 = r9
17 #1 t176 v80 x.15 |: A1 B1 :|
18 A2 = o3 |: c3 r3 r5 r7 :|4
19 B2 = r9
20 #2 t176 v80 x.15 |: A2 B2 :|
21 """
22
23 mml = MML(seq, voices=NUM_VOICES).play()
24
25 tab = CosTable([(0,0),(64,1),(1024,1),(4096, 0.5),(8191,0)])
26 envs = [TrigEnv(mml.getVoice(i), table=tab,
27                 dur=mml.getVoice(i, "dur"),
28                 mul=mml.getVoice(i, "amp"))
29         for i in range(NUM_VOICES)]
30 voices = [SineLoop(freq=mml.getVoice(i, "freq"),
31                    feedback=mml.getVoice(i, "x"),
32                    mul=envs[i]).mix(2).out()
33         for i in range(NUM_VOICES)]
34
35 s.gui(locals())
```

In Pyo's MML we can include comments inside our multi-line strings. These start with the semi-colon character, and are used in Script 11.2 in lines 8 and 9 to give information about the piece. We can also break lines with a backslash, which is done in lines 10 and 12. This way, we can conveniently separate bars so it is easy to follow when reading the code. We can also define arbitrary variables in our MML sequences with the letters x, y, and z. In Script 11.2 we define a value for the x variable, for each voice, in lines 14, 17, and 20. We access these variables through the getVoice() method, in line 31. Figure 11.1 shows the score of this music, so you can compare it to the MML notation.

Musica Ricercata III

Györgi Ligeti

Figure 11.1 The first five bars of Györgi Ligeti's *Musica Ricercata III*.

11.2.3 Tuplets and Randomness

Note tuplets are enclosed in round brackets. The duration of the tuplet is specified in the same way as with the notes. The duration of each note in a tuplet is a division between the duration of the tuplet and the number of notes in the tuplet. For example, if the duration of a tuplet with three notes is set to dotted eighth, then each note in the tuplet will last one sixteenth.

MML supports randomness in two different ways. We can either specify a random selection of notes, or a random selection of values. Randomness is indicated with the question mark. For notes, we must follow the question mark with square brackets, inside of which we must write the notes to randomly choose from. The following line will choose a random note between C, E flat, G, and B flat.

```
?[c e- g b-]
```

To make a random choice of values, we must specify the range within which the random choice will be made. This is done by following the question mark with curly brackets, inside of which we either provide two values, to specify the two extremes of the desired range, or one value, where the range will be from 0 to that value. The following line defines a random tempo between 80 and 120 BPM.

```
t?{80, 120}
```

Script 11.3 is a simple example of how tuplets and random values work. Both macros make a random choice for the octave, between 4 and 6. Then, macro A starts with a triplet of random notes, with a total duration of a quarter note, then plays a random half note, and another random quarter note. All notes in this macro are chosen between C, E, and G. Macro B starts with a quintuplet of random notes, with a duration of a half note, and then plays two random quarter notes. All notes are chosen between D, F, and A.

Script 11.3 Random notes and tuplets in MML.
```
1 from pyo import *
2
```

```
 3 NUM_VOICES = 2
 4
 5 s = Server().boot()
 6
 7 seq = """
 8 A = o?{4,6} (?[c e g] ?[c e g] ?[c e g])5 \
 9     ?[c e g]7 ?[c e g]5
10 B = o?{4,6} (?[d f a] ?[d f a] ?[d f a] ?[d f a] ?[d f a])7 \
11     |: ?[d f+ a]5 :|
12 #0 t120 v60 x.02 A B
13 BEAT = o7 |: a5 :|4
14 #1 t120 v60 x.08 BEAT
15 """
16
17 mml = MML(seq, voices=NUM_VOICES, loop=True).play()
18
19 tab = CosTable([(0,0),(64,1),(1024,1),(4096, 0.5),(8191,0)])
20 envs = [TrigEnv(mml.getVoice(i), table=tab,
21                 dur=mml.getVoice(i, "dur"),
22                 mul=mml.getVoice(i, "amp"))
23         for i in range(NUM_VOICES)]
24 voices = [SineLoop(freq=mml.getVoice(i, "freq"),
25                    feedback=mml.getVoice(i, "x"),
26                    mul=envs[i]).mix(2).out()
27          for i in range(NUM_VOICES)]
28
29 s.gui(locals())
```

The macro BEAT is created to keep the beat so we can compare the rhythm of the tuplets to a steady 4/4 beat. It is a distinct high pitch A note, with a bit sharper timbre than the first voice, as the feedback attribute of the SineLoop() oscillator that plays the beat gets the value 0.08, whereas the oscillator that plays the tuplets gets 0.02, from the x variable.

One last thing to note in this script is the loop kwarg of the MML() class. In line 17 we set it to True, so the sequences we create in the MML multi-line string will keep on looping while the Python script is running. The default value of this kwarg is False, so we can explicitly set the number of repetitions within the MML string.

The MML() class comes with a text editor of its own, which we can launch with its editor() method. As with all Pyo's GUI, you will need to have wxPython installed to use this editor. If we type mml.editor(), a window with the contents of the MML string will appear. In that window we can edit our string and hit Ctl+Return to change it in the MML() sequence. This can be helpful when prototyping sequences, so you can hear the results immediately and not having to stop and re-launch the script every time. By setting the updateAtEnd and loop attributes to True, the MML string will be updated only when the running string has finished, so we can use this feature to live-code music sequences.

As you can imagine, the MML feature of Pyo can be combined with music21 so we can see the scores of the sequences we create. Or we can load music from music21's corpus and parse it to create an MML string with its contents, so we can play back this music this way, instead of creating our own sequencer or using the Events Framework. It is beyond the scope of this chapter to make this connection, so this is left to the discretion of the reader.

Documentation Pages:

MML API: https://belangeo.github.io/pyo/api/classes/mmlmusic.html
MML(): https://belangeo.github.io/pyo/api/classes/mmlmusic.html#mml

11.3   What is the Prefix Expression

Pyo includes another small programming language, based on prefix notation, also known as Polish notation. This notation was invented by the Polish logician and philosopher Jan Łukasiewicz (Hamblin, 1962). As its name implies, the operators in this notation are written before their operands, in contrast to infix notation, where operators are written in between their operands. For example, in prefix notation, an addition of two numbers is written as in the line below.

```
+ 2 3
```

In Pyo's mini-language that employs prefix notation, every operation must be placed inside round brackets. So the entire line above must be enclosed in round brackets when used within Pyo. In prefix notation, and Pyo's prefix mini-language, nesting is used to achieve more complex operations. To add a multiplication to be performed after the addition of the line above, we must write the line below.

```
(* (+ 2 3) 10)
```

This line adds 2 and 3, and then multiplies the result by 10. As with the MML, expressions in this language are written in multi-line strings, and are then fed to a Pyo class that parses them and produces the output. This class is Expr(). An advantage this language has over writing our own functions or classes is that the operations written in it are performed at sample-level. If we need access to specific samples within a sample block, like every previous sample, we can do this with this prefix mini-language. On the contrary, it is not possible to have access to specific samples with user-defined classes, because all PyoObjects compute their samples internally, in blocks, and output vectors of samples, instead of outputting samples one by one. In the sections below we will see scenarios where this feature can be useful.

11.4   The Prefix Expression Language

Even though, small, the prefix expression language covers a wide range of operations, from arithmetic and trigonometric functions, to conditional operations and periodic functions. It even enables users to define their own functions within expressions, use local and global variables, as well as external input.

11.4.1   Arithmetic Operators

Pyo's prefix language includes the following arithmetic operators: +, -, *, /, ^, %, neg. The first four are obvious as to what they do. The caret symbol (^) raises a value to a power. The modulo operator (%) returns the floating point remainder of the division of two values, and the neg operator returns the negative of a value. All operators but the last, take two operands, like the addition line above, and the neg operator takes only one, like the line below.

```
(neg x)
```

11.4.2   *Moving Phase Operators*

Moving phase operators increment or decrement a value within a given range. The following operators are available: ++, –, ~. All operators take two operands. The first two increment or decrement respectively, the value of their first operand and wrap it around 0 and the value of their second operand. For example, the following line increments x by 1, and wraps its value around 0 and 10.

```
(++ x 10)
```

The tilde operator creates a periodic ramp between 0 and 1 with the frequency defined by the first operand and the phase defined by the second. The phase can be omitted, so this operator can also take one argument only. This periodic ramp is equivalent to a Phasor(). The following line creates a ramp between 0 and 1 with frequency 440.

```
(~ 440)
```

11.4.3   *Conditional Operators*

Conditional operators include tests against values that determine whether they are greater than, equal to, smaller than, or different from another value, if tests, and Boolean operations. The following operators are available: <, <=, >, >=, ==, !=, if, and, or. All but the if operator take two operands. The first four are obvious as to what they do. The double equal signs tests for equality, and the != operator tests for inequality. The and operator is a Boolean AND, and the or is a Boolean OR. All these operators return 1 if their test is successful, otherwise they return 0. The if operator is used with the following structure:

```
(if (condition) (then) (else))
```

If the condition is True, then this operator returns its then value. If the condition is False, it returns its else value. The following line tests if x is greater than y, and if this is True, it returns the difference of these two operands, otherwise it returns their sum.

```
(if (> x y) (- x y) (+ x y)).
```

11.4.4   *Trigonometric Functions*

The following trigonometric functions are available in Pyo's prefix language: sin, cos, tan, tanh, atan, atan2. All but atan2 take one operand. The tan function returns the tangent of the radians expressed by its operand. The tanh function returns the hyperbolic tangent of the radians expressed by its operand. The atan function returns the principal value of the arc tangent of the radians expressed by its operand, and atan2 returns the principal value of the arc tangent of the radians expressed by the division between its two operands.

11.4.5   *Power and Logarithmic Functions*

The following functions are available: sqrt, log, log2, log10, pow. All but pow take one operand. The log function returns the natural logarithm of its operand, that is, the base-e logarithm, where e is Euler's number, approximated to 2.718281828459045. The log2 function returns the base-2 logarithm of its operand, and `log10` returns the base-10 logarithm of its operand. The `pow` function raises its first operand to the power of its second operand, so the following line will raise x to the y power.

```
(pow x y)
```

11.4.6 Clipping Functions

The following clipping functions are available: `abs`, `floor`, `ceil`, `exp`, `round`, `min`, `max`, `wrap`. All but `min` and `max` take one operand. The `min` and `max` functions take two operands. These output the minimum and maximum value respectively, between their two operands. The `abs` function returns the absolute value of its operand. The `floor` function returns the largest integral value that is not greater than its operand. This is essentially like trimming out the decimal numbers of a float, so if the operand is 10.4, the output will be 10. The `ceil` function returns the smallest integral value that is not less than its operand, rounding the operand upward. For 10.4, `ceil`'s output would be 11. The `exp` function raises the constant value e to the power of its operand. `round` rounds a float value to its nearest integer, and `wrap` wraps its operand between 0 and 1.

11.4.7 Random Functions

Pyo's prefix language provides two random functions, `randf` and `randi`. They both take two operands which define the range between which the functions will produce a random number. The former outputs floats, while the latter outputs integers.

11.4.8 Complex Numbers

With the prefix language we can define complex numbers, or retrieve either the real or the imaginary part. The `complex` operator takes two operands and returns a complex number where the first operand is the real part and the second is the imaginary part. To retrieve the real part from a complex number we can use the `real` operator which takes one operand that must be a complex number. To retrieve the imaginary part we can use the `imag` operator in a similar way. The exemplar line below returns the real part of the complex number defined by x and y. This line essentially returns x.

```
(real (complex x y))
```

11.4.9 Filter Functions

Pyo's prefix language provides a set of raw filters that can be used to construct more high-level, complete filters. These are the following: `delay`, `sah`, `rpole`, `rzero`, `cpole`, `czero`. The `delay` filter takes one operand. It is a one-sample delay of this operand. `sah` is a sample-and-hold filter that takes two operands. It will sample one value of its first operand and will hold it, as long as its second operand does not drop its value. For example, the line below provides white noise as the first operand to `sah` and a ramp with a 1Hz frequency as the second operand.

```
(sah (randf -1 1) (~ 1))
```

This line results in a random value being sampled every one second. This happens because the (~ 1) part of this expression creates a rising ramp from 0 to 1 that resets its phase every one second. When this phase resets, the output of this function will be less than the previous output, causing `sah` to sample a new value of its first operand, whereas while the ramp is rising, every sample is greater than its previous one, not triggering `sah`.

The `rpole` filter is a real one-pole recursive filter. It takes two operands and returns the following (for operands x and y): `x + last_output * y`. The `rzero` filter is a real one-zero non-recursive filter. Its output is: `x + last_x * y`. The `cpole` filter is a complex one-pole recursive filter. It takes two operands which both must be complex. The first must be a complex signal that will be filtered, and the second a complex coefficient. The `czero` filter is a complex one-zero non-recursive filter. Like `cpole`, both its operands must be complex.

To be able to use these raw filters, one needs some theoretical background in filter design. An explanation of how we can use the `rpole` filter to create a first-order Infinite Impulse Response (IIR) lowpass filter is provided in this section. This is taken from Pyo's documentation and broken to smaller chunks to make it easier to understand. Shell 11.1 is a Python version of what the second operand of this filter must be.

Shell 11.1 Calculating the coefficient for a first-order lowpass filter.

```
>>> import math
>>> cutoff = 1000
>>> coeff = pow(
...     math.e, ((-(math.pi*2) * cutoff)/s.getSamplingRate())
... )
```

Shell 11.1 assumes that we have imported Pyo and have booted the server, so we can get the sampling rate at the end of the fourth line. The `cutoff` variable defines the cutoff frequency of the filter. In the prefix language, the fourth line of Shell 11.1 is expressed by the line below.

```
(exp (/ (* (neg twopi) 1000) sr))
```

In this line, we hard-code the cutoff frequency to 1000. It is possible to use variables with this language, but we haven't discussed this feature yet, so, for the sake of simplicity, we hard-code it. The `sr` is a prefix language constant, returning the sampling rate. We will see all the available constants later on. Let's assume that we store the output of the line above to a variable called `#coeff` (the hash symbol is used in the prefix language syntax to define variables, but more on that later on). To conclude our filter, we will pass white noise as the first operand, which is the signal to be filtered by `rpole`. This signal will be multiplied by 1 − #coeff (in infix notation) before it is processed. The line below concludes our filter in the prefix language.

```
(rpole (* (randf -1 1) (- 1 #coef)) #coef)
```

11.4.10   Multiple Outputs

By default, an expression written in the prefix language will produce a single audio stream, whose output will be the last line in the multi-line string containing the expression. The `out` operator enables us to output more than one channel. It takes two operands, where the first defines the channel, and the second defines the signal to be written to that channel.

11.4.11   *Constants*

The prefix language offers a few constants that facilitate some operations. These are `pi`, `twopi`, `e`, and `sr`. We have already seen the last one. The `e` constant returns Euler's e value. The first two are obvious as to what value they return. It is also possible to create user defined constants by using the `const` operator. It takes one operand, which is the constant value to return.

11.4.12   *Accessing Input and Output Samples*

Since the prefix language expressions function at a one-sample basis, it is useful to have an easy way to access any sample we want in the audio buffer. The notation to access input or output samples is this: `$xc[n]` for the n input sample of the c audio stream, `$yc[n]` for the n output sample of the c audio stream. The number to replace c should be a zero-based channel count, where 0 is the first audio stream, 1 is the second, and so on. For the first audio stream, this value can be omitted. This means that `$x[0]` is the same as `$x0[0]`. This expression returns the current sample of the first audio input stream. To get the last output sample of the second audio stream, we must type `$y1[-1]`. The maximum index that we can use to access samples is 0, and all past samples are indexed with negative numbers.

11.4.13   *Defining Functions Within the Prefix Language*

It is possible to define our own functions within the prefix language. The syntax for defining a function is shown in the line below.

```
(define funcname (body))
```

The keyword `define` is used to tell the language evaluator that what follows is a function definition. Then we use a name for our function, and then we enclose the body of the function in round brackets. As with the rest of the prefix language, the whole function definition, including the function name and the `define` keyword, must be enclosed in round brackets. Script 11.4 shows a simple function definition. In this script, we define a clipping function using the `min` and `max` operators of the prefix language. We then call the function in line 5 with three arguments, the value we want to clip, the lower bound, and the upper bound. In the function body we can access the arguments with `$1` for the first argument, `$2` for the second, and so on. The value to be clipped can be passed via PyoObjects. We will see later on how this is possible.

Script 11.4 A function definition in the prefix language.
```
1 (define clip (
2    (min (max $1 $2) $3)
3    )
4 )
5 (clip $x[0] 5 10)
```

11.4.14   *State Variables*

We can create state variables with the keyword `let`. In the previous example with the low pass filter, we supposed that we stored the first line to a variable called `#coeff`. This is indeed possible as it is shown in Script 11.5.

Script 11.5 Using a state variable to store the lowpass filter coefficient calculation.

```
1 (let #coef (exp (/ (* (neg twopi) 1000) sr)))
2 (rpole (* (randf -1 1) (- 1 #coef)) #coef)
```

The #coeff state variable in Script 11.5 is global to the prefix expression. We can define local variables inside functions, which are only accessible within the function. If a global and a local variable share the same name, the local variable will be chosen. Script 11.6 defines a function for our lowpass filter and defines the #coeff variable twice, once in the global scope and once in the local scope of the function. The local one that has a lower hard-coded cutoff frequency is the one that is used.

Script 11.6 Global vs local state variables in the prefix language.

```
1 (let #coef (exp (/ (* (neg twopi) 500) sr)))
2 (define lowpass (
3     (let #coef (exp (/ (* (neg twopi) $1) sr)))
4     (rpole (* (randf -1 1) (- 1 #coef)) #coef)
5     )
6 )
7 (lowpass 100)
```

State variables can also be used to create a one-sample feedback, if they are used inside their own definition. Script 11.7 creates a phasor by incrementing and wrapping a variable inside its own definition. The variable is thus used before it is actually created. In this script we can also see a comment in the prefix language, which is the same as comments in C++, starting with two forward slashes.

Script 11.7 Using a state variable inside its own definition.

```
1 (define phasor (
2         (let #ph
3             (wrap (+ #ph 0.01)) // increment before end of def
4         )
5         #ph
6     )
7 )
8 (phasor)
```

11.4.15   User Variables

User variables are different to state variables in that they can be modified from outside of the prefix language, by the Python script. Their syntax is similar to that of state variables, but their keyword is var. If we go back to our lowpass filter expression, we can set the cutoff frequency to an initial value, but make it so that it can be changed while the expressions are running. Script 11.8 shows this.

Ingredients:

- Python3
- The Pyo module
- The wxPython module, if you want to use Pyo's GUI

- A text editor (preferably an IDE with Python support)
- A terminal window in case your editor does not launch Python scripts

Process:

In this script we can finally see how we can use the prefix language inside a Python script. We will first discuss the prefix language expression and then how we can evaluate it and send signals and get its signal inside the script. In line 6 we define a user variable named #cutoff, and initialise it to 100. We use this variable inside our lowpass function, in line 8. In line 12 we call this function and pass the current input sample we get from a PyoObject. In the function body, we access this input sample with $1 as we would access any argument to a function in the prefix language. This argument has replaced the (randf -1 1) part in line 4 of Script 11.6.

Script 11.8 Defining a user variable in the prefix language.

```
1 from pyo import *
2
3 s = Server().boot()
4
5 expression = """
6 (var #cutoff 100)
7 (define lowpass (
8      (let #coef (exp (/ (* (neg twopi) #cutoff) sr)))
9      (rpole (* $1 (- 1 #coef)) #coef)
10      )
11 )
12 (lowpass $x[0])
13 """
14
15 expr = Expr(Noise(), expression)
16 sp = Spectrum(expr)
17 mix = Mix(expr, voices=2).out()
18
19 s.gui(locals())
```

Once we define our prefix expression, we can evaluate it with the Expr() class. This class takes a PyoObject or a list of PyoObjects in its first argument. The audio stream of this object is passed to the prefix expression and it is accessible with $x[0], in line 12. In Script 11.8 we pass white noise to our expression. The second argument of the Expr() class is the expression we want to evaluate.

The output of the Expr() class is either the last line of the expression it evaluates, or any out operator we might use in our expression. The output of this class is a Pyo audio stream and it can interact with the entire Pyo ecosystem. To change a user variable, we need to call the setVar() method of the Expr() class, with the name of the variable as a string in its first argument, and the value we want to pass to this variable in its second argument. For Script 11.8, to change the cutoff frequency of the lowpass filter we need to type the following line, replacing the number 500 to any frequency we want to set the filter's cutoff to. If you launch this script, you can call this method from the interpreter entry of Pyo's server window.

```
expr.setVar("#cutoff", 500)
```

11.4.16   Loading External Files

In programming, it is often a good idea to abstract our code to functions and classes, and many times we want to save code that we often use, in a file so we can recall it in many different projects. The same applies to the prefix language, where it is possible to load a file with prefix expressions in the main Python file where we evaluate an expression. This is possible with the load operator.

Ingredients:

- Python3
- The Pyo module
- The wxPython module, if you want to use Pyo's GUI
- A text editor (preferably an IDE with Python support)
- A terminal window in case your editor does not launch Python scripts

Process:

We can write some oscillator functions in one file, and some filter functions in another file, and load both files to a Python script so we don't have to re-write these functions, plus we can minimise the length of our prefix expression. Script 11.9 shows a function for a square wave oscillator, and Script 11.10 shows the lowpass filter function. In the latter, the #cutoff variable has been replaced by an argument, accessed with $2.

Script 11.9 A square wave oscillator in the generators.expr file.
```
1 (define square (
2      (if (< (~ $1) 0.5) (1) (-1))
3      )
4 )
```

Script 11.10 A lowpass filter in the filters.expr file.
```
1 (define lowpass (
2      (let #coef (exp (/ (* (neg twopi) $2) sr)))
3      (rpole (* $1 (- 1 #coef)) #coef)
4      )
5 )
```

In Script 11.11 we load the two files in lines 6 and 7, named generators.expr and filters.expr respectively (you can choose any suffix for the files you want), and we call the functions of scripts 11.9 and 11,10 in line 10, without needing to re-define these two functions. Note that we pass a Sig(0) to the first argument of the Expr() class, because we don't need to process any signal passed to our expression. Once we launch the script, we can call the setVar() method of the Expr() class to change any of the two user variables defined in lines 8 and 9. This example is inspired by Pyo's documentation on the Prefix Expression language.

Script 11.11 Loading files in the prefix language.
```
1 from pyo import *
2
3 s = Server().boot()
```

```
 4
 5 expression = """
 6 (load generators.expr)
 7 (load filters.expr)
 8 (var #freq 200)
 9 (var #cutoff 100)
10 (lowpass (square #freq) #cutoff)
11 """
12
13 expr = Expr(Sig(0), expression)
14 sc = Scope(expr)
15 mix = Mix(expr, voices=2).out()
16
17 s.gui(locals())
```

11.5   A few more Examples with Prefix Expressions

The prefix language can be very useful in various use cases, like when we need to define a function where sample accuracy is important, or when we want to be able to edit expressions on-the-fly.

Ingredients:

- Python3
- The Pyo module
- The wxPython module, if you want to use Pyo's GUI
- A text editor (preferably an IDE with Python support)
- A terminal window in case your editor does not launch Python scripts

Process:

Script 11.12 defines a triangle wave oscillator that feeds if output back to its own phase, like the SineLoop() class does. It is possible to write such a class in Python, with PyoObjects, but we would not get the same brilliance we get with the prefix language, because if we defined this oscillator in Python, the feedback of the oscillator would occur with one sample-block latency. This means that the first sample we would be feeding back to the phase of the oscillator, would be the first sample of the previous sample block. With the prefix language, we can feed back the last sample of the current sample block, and get the desired brilliance we would expect from a feedback algorithm.

Script 11.12 A triangle wave feedback oscillator in the prefix expression language.

```
 1 from pyo import *
 2
 3 s = Server().boot()
 4
 5 expression = """
 6 (var #freq 440)
 7 (var #fb 0.1)
 8 (define triangle (
 9          (let #ph (wrap (+ (~ #freq) (* $1 #fb))))
```

```
10                  (- (* (min #ph (- 1 #ph)) 4) 1)
11      )
12 )
13 (triangle $y[-1])
14 """
15
16 expr = Expr(Sig(0), expression, mul=0.5)
17 sc = Scope(expr)
18
19 mix = Mix(expr, voices=2).out()
20
21 s.gui(locals())
```

Note that it is not possible to have access to samples inside a function, so we pass the last output sample as an argument, in line 13, and get access to it inside the function with $1, in line 9. In this script, we have set a value to the `mul` attribute of the `Expr()` class, an attribute that is common among almost all Pyo classes. Once you launch this script, you can call the `setVar()` method and change the two user variables defined in lines 6 and 7 through the interpreter entry of Pyo's server window.

A feature of the prefix expression language we haven't seen in an example yet is the multiple outputs. This is shown in Script 11.13. We first create a Ring Modulation (RM) output, by multiplying the current samples of the first two input streams. Then we output the RM out the first channel, and an Amplitude Modulated (AM) version of the RM out the second channel. All three input streams to the expression must be bipolar, as we take care to unipolarise the Amplitude Modulator in line 8. To be able to smoothly change between different inputs, we use the `InputFader()` class, in line 16. We pass two `rm_in*` objects and the `am_in` object as a list to its input. In the `Expr()` class, we need to define the number of outputs by passing the correct value to its `outs` kwarg.

Script 11.13 Multiple outputs with Expr() and the prefix expression language.

```
 1 from pyo import *
 2
 3 s = Server().boot()
 4
 5 expression = """
 6 (let #rm (* $x0[0] $x1[0])) // Ring Modulation
 7 (out 0 #rm)
 8 (out 1 (* #rm (+ (* $x2[0] 0.5) 0.5)))
 9 """
10
11 rm_in1 = Sine(freq=200)
12 rm_in2 = SineLoop(freq=52.7, feedback=0.08)
13 rm_in3 = RCOsc(freq=180)
14 rm_in4 = SuperSaw(freq=53.2)
15 am_in = LFO(freq=.5, type=3)
16 expr_in = InputFader([rm_in1, rm_in2, am_in])
17 expr = Expr(expr_in, expression, outs=2, mul=0.5).out()
18 sc = Scope(expr)
19
20 s.gui(locals())
```

Launch this script and use the interpreter entry in Pyo's server window to change the input to the `expr_in` object, by calling its `setInput()` method, like in the line below. Set the time you want in seconds, to cross-fade between the old and new input, with the `fadetime` kwarg.

```
expr_in.setInput([rm_in3, rm_in4, am_in], fadetime=2)
```

As with the `MML()` class, the `Expr()` class can launch its own editor window, by calling its `editor()` method. A simple text editor will pop up, where the user can edit the prefix expression and update it by hitting Ctl+Return. To use this feature, you will need to have wxPython installed. As with `MML()`'s editor, using this feature can be useful for refining your prefix expressions while prototyping, to get the results you want. It can also be useful in a live coding context, but do expect audible clicks while updating your expressions, because there is no cross-fade happening between expression evaluations, and most likely the audio streams will jump from their old values to their new ones.

Documentation Pages:

Prefix Expression: https://belangeo.github.io/pyo/api/classes/expression.html
`Expr()`: https://belangeo.github.io/pyo/api/classes/expression.html#expr
`InputFader()`: https://belangeo.github.io/pyo/api/classes/internals.html#inputfader
Prefix Expression Tutorial: https://belangeo.github.io/pyo/examples/23-expression/index.html

11.6 Conclusion

In addition to the Events Framework, and the ability to write our own classes with PyoObjects, Pyo includes two mini-languages for music composition and sample-accurate algorihm coding, MML and the prefix expression language. The former enables users to write music segments in a language with a Western-music mindset that uses jargon that is understandable by composers familiar with it. The evaluator class of this language, `MML()`, provides all the necessary signals and information to realise the compositions written in it, enabling users to play their music sequences with a handful of lines of code.

The prefix expression language might seem a bit more esoteric, since infix notation is the standard that most people are used to. Once you get the hang of it though, it becomes easy to prototype algorithms of various sorts, very fast. This language enables users to go down to sample-level accuracy, something that is desirable in cases where a single-sample feedback is required. This level of sample accuracy might not possible with standard Python classes, and without the prefix expression language, users would likely have to write their own Pyo classes in C, when such accuracy is needed.

Both the `MML()` and the `Expr()` classes provide their own editor where the user can change the strings that are evaluated by the respective class. This feature provides a very flexible framework for prototyping, as users don't need to quit and re-launch their scripts when they need to change their sequences or algorithms. Being able to edit sequences and prefix expressions on-the-fly enables live coding too, but in the case of the prefix language, audible clicks should be expected, as changes in the algorithm will most likely result in sudden changes in the values of the audio streams. With MML though, it is possible to live code sequences without audible clicks, as we can set the class to update the running sequence only when finished.

11.7   Exercises

11.7.1   *Exercise 1*

Write a few more bars of Ligeti's *Musica Ricercata III* in MML.

Tip: Finding the sheet music should not be very difficult, even if it is in the form of an online video. Compare the sound result to the score, to make sure you have written the music correctly.

11.7.2   *Exercise 2*

Write a function for a square wave oscillator with duty cycle control in the prefix expression language. Save it in the external file used in Script 11.11 and test it with this script.

11.7.3   *Exercise 3*

Create a two-voice phase shifting melody with MML, in the style of Steve Reich.

Bibliography

Hamblin, C.L. (1962) 'Translation to and from Polish Notation', *The Computer Journal*, 5(3), pp. 210–213. Available at: https://doi.org/10.1093/comjnl/5.3.210.

Matsushima, T. (1997) 'Music Macro Language', in *Beyond MIDI: The Handbook of Musical Codes*. Cambridge, MA, USA: MIT Press, pp. 143–145.

12 Writing Your Own GUI

Even though Pyo comes with its own GUI that works very well, there might be some cases where we need to write our own GUI widgets. One such case can be an entry to type numeric values to be passed to an attribute of a PyoObject, instead of calling a function with an argument. Another case could be the need for a selection list, so we can easily select the waveform of an oscillator, or some other feature with many options. A more obvious case is using Pyo with a Python version for which wxPython has no wheels (yet). Pyo being a Python module, can be combined with any existing module for GUI creation.

What You Will Learn

After you have read this chapter you will:

- Know how to prototype simple GUI widgets in two Python modules
- Know how to control parameters of PyoObjects with your custom widgets
- Be able to cover minor gaps in Pyo's own GUI widgets

12.1   Available Python Modules for GUI Widgets

There are plenty of modules in Python for designing custom GUI widgets, including wxPython, that is natively used by Pyo. In this chapter we will look at two modules. The first one is tkinter, Python's native GUI module. This module is an interface to Tcl/Tk, a GUI toolkit. The second module we will look at is PyQt, a Python bindings module to the Qt toolkit. In this chapter we will create minimal user interfaces with these modules, to cover the few aspects that Pyo's native GUI widgets don't cover. We will create an interface with each module for the same functionality. This interface will include a tickbox to enable/disable the DSP, a slider to set the frequency of an oscillator of the `LFO()` class, a text entry where the user can type the frequency value, and a radio with eight buttons, to set the waveform of the oscillator.

12.2   Using Python's Native Tkinter

Tkinter is Python's native module for programming GUI widgets, so we don't need to install anything to complete this section. It is an interface to Tcl/Tk, an open-source toolkit that was initially developed in the 1990s. Tcl/Tk is used by many applications, but with the advent

DOI: 10.4324/9781003386964-12

of other toolkits, like Qt, its popularity has somewhat declined. Nevertheless, it is still actively maintained, and since Python provides a native module for it, we will build our project with it.

Ingredients:

- Python3
- The Pyo module
- A text editor (preferably an IDE with Python support)
- A terminal window in case your editor does not launch Python Scripts

Process:

The code for the interface of this project is shown in Script 12.1. We will look at each part of the interface separately as we read through the code. Figure 12.1 shows the resulting interface.

Script 12.1 The tkinter version of the custom GUI interface.

```
 1 import tkinter as tk
 2 import pyo
 3
 4 s = pyo.Server().boot()
 5
 6 sig = pyo.SigTo(200)
 7 osc = pyo.LFO(freq=sig, mul=.2)
 8 mix = pyo.Mix(osc, voices=2).out()
 9
10 class TkWidgets(tk.Frame):
11     def __init__(self, master):
12         tk.Frame.__init__(self, master)
13         self.tickbox_var = tk.IntVar()
14         self.tickbox = tk.Checkbutton(
15                     master, text="DSP",
16                     variable=self.tickbox_var,
17                     command=self.tickbox_action,
18                     font=("Liberation Mono", 10)
19                 )
20         self.tickbox.place(x=50, y=20)
21
22         self.freq_label=tk.Label(
23                     master,text="Frequency",
24                     font=("Liberation Mono",15)
25                 )
26         self.freq_label.place(x=50, y=70)
27
28         self.slider = tk.Scale(
29                     master, from_=0.00, to=300.0,
30                     orient=tk.HORIZONTAL,
31                     length=300,
32                     tickinterval=0,resolution=0.01,
33                     font=("Liberation Mono",10),
34                     command=self.set_freq
```

```
35                          )
36          self.slider.set(200)
37          self.slider.place(x=50, y=100)
38
39          self.entry = tk.Entry(
40                      master, font=("Liberation Mono", 12)
41                      )
42          self.entry.place(x=50, y=150)
43          self.entry.bind('<Return>', self.get_freq)
44
45          self.radio_label = tk.Label(
46                      master, text=" Waveform ",
47                      font=("Liberation Mono", 20)
48                      )
49          self.radio_label.place(x=50, y=200)
50          self.waveforms = ["Saw up          ",
51                          "Saw down        ",
52                          "Square          ",
53                          "Triangle        ",
54                          "Pulse           ",
55                          "Bipolar Pulse   ",
56                          "Sample and Hold",
57                          "Modulated Sine "]
58          self.radio_var = tk.IntVar()
59          self.radio=[tk.Radiobutton(
60                      master,
61                      text=self.waveforms[i],
62                      variable=self.radio_var,
63                      value=i,
64                      font=("Liberation Mono", 10),
65                      command=self.radio_func).place(
66                        x=50,y=210+((i+1)*20)
67                      ) for i in range(len(self.waveforms))]
68
69  ############## Tick box ###############
70  def tickbox_action(self):
71      if self.tickbox_var.get()== 1:
72          s.start()
73      else:
74          s.stop()
75
76  ############### Slider ################
77  def set_freq(self, value):
78      sig.setValue(float(value))
79
80  ############# Number entry ############
81  def get_freq(self, arg):
82      sig.setValue(float(self.entry.get()))
83      self.entry.delete(0, tk.END)
84
85  ################ Radio ################
86  def radio_func(self):
```

```
87                 osc.setType(self.radio_var.get())
88
89
90 if __name__ == "__main__":
91     root = tk.Tk()
92     root.configure(background='light grey')
93     root.geometry('400x440')
94     TkWidgets(root)
95     root.mainloop()
```

In the first lines of this script we import the two modules we will use, and then we initialise all PyoObjects. Since we will not be using Pyo's widgets, we should start the Pyo server by calling s.start(). We don't do that though, as we want to be able to start and stop it through the widgets we will build.

All tkinter widgets are placed inside a class called TkWidgets(). This class inherits from the Frame() class of the tkinter module. The master argument passed to our class is essentially a Tk() class that is passed in line 94. This is needed in every widget of this module. The first widget we create is the tickbox, labelled "DSP". This is done in lines 13 to 20. We set a variable to hold its value, and pass a function that will be called when we click on it. This function is defined in line 70, and starts or stops the Pyo server, depending on the value of the tickbox. We use the *Liberation Mono* font, because we need a monospace font in our radio – the last widget – so, for visual integrity, we use this font in all widgets. In line 20 we place the widget inside the layout of our program, by defining the X and Y coordinates, with (0, 0) being at the top left corner of the window.

The next widget we create is a label with the text "Frequency", which we use to label the slider and text entry of our program. Again, we use the *Liberation Mono* font. In lines 28 to 37 we create the slider. There we set its range, its orientation, its resolution, its length, and the function it will call when its value changes. When calling this function, this object will pass its value implicitly, without us having to specify this, but we need to set the correct number of arguments to the function definition, in line 77. We initialise the slider to the value 200, and we place it in our layout.

Lines 39 to 43 create a text entry, where we can type a frequency value. In line 43 we bind the Return key to the get_freq() function. This function will be called with an argument, which is a tkinter.Event() class, passed implicitly. We need to define our function with this argument, but we are not actually using it. What we need, is the actual text of this entry, which we retrieve with self.entry.get(), in line 82. We convert this to a float with Python's float() class. In case what we type in the entry is not a float, we will get a ValueError in our traceback, but our program will not exit. We could use a try/except block, but since our program will keep on running, this ValueError is not a problem. Once we pass the float to our oscillator, we clear the text, in line 83.

The last widget is the radio, where we can set the waveform of our oscillator. We first create another label with the text "Waveforms", to state what this radio does, and then we create a list of Radiobutton() objects, where each object takes one item of the waveforms list. Note that we make sure that each item in the list has the same number of characters, by placing as many white spaces as we need. This is done because in the layout, there is a lighter grey bounding box where the text is (see Figure 12.1). To keep the radio items aligned, we make their text lengths equal, otherwise, the bounding box of each item would have a different width. For the same reason we add two white spaces at the beginning and two at the end of the string " Waveforms ", in line 46.

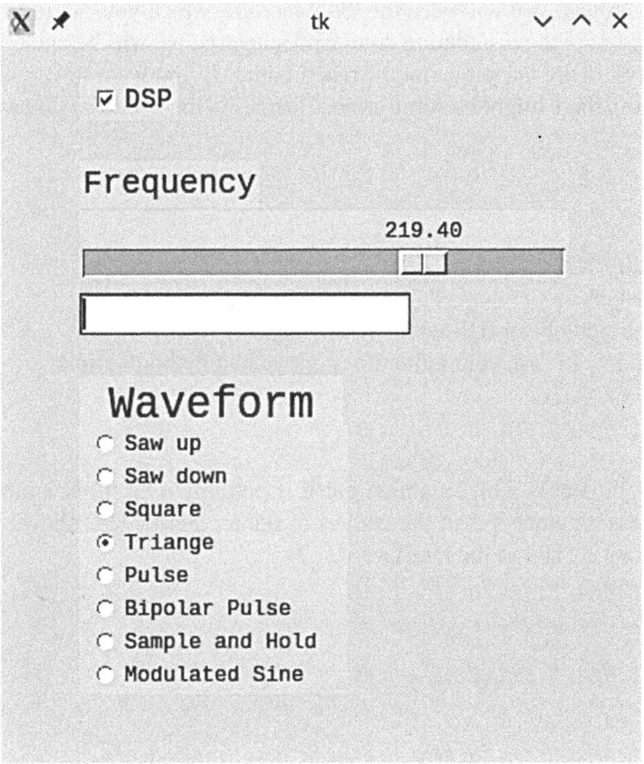

Figure 12.1 The tkinter interface.

In line 58 we create a variable that will hold each item or the radio, so we can assign the correct waveform to our oscillator. This value is the index of the corresponding radio in the list. This is done by passing i to the `value` kwarg, in line 63. Then we create the radio with its various kwargs. The `radio_func()` function is called whenever we click on a radio item, and the value of that item is passed to the `self.radio_var` variable.

In the `if __name__ == "__main__"` body of our program, we create a `Tk()` object, we set its background colour and its geometry, and then we call our `TkWidgets()` class. Then we call the `mainloop()` method of the `Tk()` class, which keeps our Script alive. If you run this Script, you should see a window like the one in Figure 12.1, where you can control the oscillator's frequency by moving the slider or by typing values in the text entry and hitting Return, and its waveform with the radio buttons. To make sound, you need to tick the "DSP" tickbox, and to stop the sound, you need to untick it.

12.3 Using PyQt

The next module we will look at is PyQt, a Python bindings module for Qt, a popular, cross-platform GUI toolkit. The PyQt module includes many classes, where each class has many methods, to cover a wide range of applications. One can go very deep in learning it, but in this chapter, we will cover the basics, building an interface with the same functionality as the one we built with tkinter.

To realise this project, you will need the PyQt module, which you can install by typing `pip install pyqt6`.[1] As the installation command suggests, we will be using PyQt6, which, at the time or writing, is the latest version of this module. If you want to use older versions (or newer in the future), there might be some minor changes in the code that you will need to make.

Ingredients:

- Python3
- The Pyo module
- The PyQt6 module
- A text editor (preferably an IDE with Python support)
- A terminal window in case your editor does not launch Python Scripts

Process:

The code for this project is a bit lengthier, but it is possibly a bit more intuitive, as the PyQt classes take far less arguments than the classes of tkinter. Script 12.2 shows the code for this project, and Figure 12.2 shows the interface.

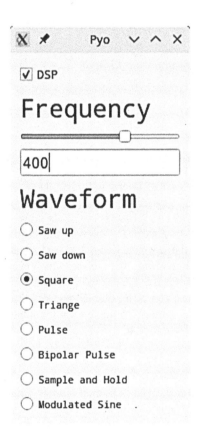

Figure 12.2 The PyQt interface.

Script 12.2 The PyQt version of the custom GUI interface.

```
 1 import sys
 2 from PyQt6.QtCore import *
 3 from PyQt6.QtGui import *
 4 from PyQt6.QtWidgets import *
 5 import pyo
 6
 7 s = pyo.Server().boot()
 8
 9 sig = pyo.SigTo(200)
10 osc = pyo.LFO(freq=sig, mul=.2)
11 mix = pyo.Mix(osc, voices=2).out()
12
13 class PyoWidgets(QWidget):
14     def __init__(self):
15         super().__init__()
16
17         fontid=QFontDatabase.addApplicationFont(
18             "DroidSansMono.ttf"
19             )
20         if fontid < 0:
21             print("Error loading font, exiting . . . ")
22             exit()
23         fontstr = QFontDatabase.applicationFontFamilies(
24             fontid
25             )
26
27         self.layout = QVBoxLayout()
28
29         self.waveforms = ["Saw up",
30                           "Saw down",
31                           "Square",
32                           "Triangle",
33                           "Pulse",
34                           "Bipolar Pulse",
35                           "Sample and Hold",
36                           "Modulated Sine"]
37
38         self.tickbox = QCheckBox("DSP")
39         self.tickbox.setChecked(False)
40         self.tickbox.stateChanged.connect(self.tickbox_state)
41         self.tickbox.setFont(QFont(fontstr[0], 10))
42         self.layout.addWidget(self.tickbox)
43
44         self.freqlabel = QLabel("Frequency")
45         self.freqlabel.setAlignment(
46           Qt.AlignmentFlag.AlignLeft
47           )
48         self.freqlabel.setFont(QFont(fontstr[0], 20))
49         self.layout.addWidget(self.freqlabel)
50
51         self.slider = QSlider(Qt.Orientation.Horizontal)
```

```
52              self.slider.setMinimum(0)
53              self.slider.setMaximum(300)
54              self.slider.setValue(200)
55              self.layout.addWidget(self.slider)
56              self.slider.valueChanged.connect(self.valuechange)
57
58              self.freqentry = QLineEdit()
59              self.freqentry.setPlaceholderText("Frequency")
60              self.freqentry.returnPressed.connect(self.freqset)
61              self.freqentry.setFont(QFont(fontstr[0], 12))
62              self.layout.addWidget(self.freqentry)
63
64              self.wavelabel = QLabel("Waveform")
65              self.wavelabel.setAlignment(
66                Qt.AlignmentFlag.AlignLeft
67              )
68              self.wavelabel.setFont(QFont(fontstr[0], 20))
69              self.layout.addWidget(self.wavelabel)
70
71              self.radio = [QRadioButton(i)
72                            for i in self.waveforms]
73              for i in range(len(self.radio)):
74                  self.radio[i].setChecked(i == 0)
75                  self.radio[i].index = i
76                  self.radio[i].toggled.connect(self.set_waveform)
77                  self.radio[i].setFont(QFont(fontstr[0], 8))
78                  self.layout.addWidget(self.radio[i])
79
80              self.setLayout(self.layout)
81              self.setWindowTitle("Pyo")
82
83      def valuechange(self):
84          freq = self.slider.value()
85          sig.setValue(freq)
86
87      def freqset(self):
88          freq = float(self.freqentry.text())
89          self.freqentry.clear()
90          sig.setValue(freq)
91
92      def tickbox_state(self):
93          if self.tickbox.isChecked():
94              s.start()
95          else:
96              s.stop()
97
98      def set_waveform(self):
99          for r in self.radio:
100             if r.isChecked():
101                 osc.setType(r.index)
102                 break
103
```

```
104
105 if __name__ == '__main__':
106     app = QApplication(sys.argv)
107     widgets = PyoWidgets()
108     widgets.show()
109     sys.exit(app.exec())
```

The first thing we do is load a font so we can easily control the font size in the various widgets. This is done in lines 17 to 19. The font file has to be in the same directory as the Python Script. You can use any font you like. Script 12.2 uses a monospace font, but that is not necessary, as we are not facing the same issue we did with tkinter, where the radio button labels would have different bounding box widths. Once we successfully load our font, we get the string of the font from PyQt in line 23.

In line 27 we create our layout. In line 29 we create the list with the waveform names. We don't need to add white spaces in this version, since, as you see in Figure 12.2, there is no bounding box around the radio button labels. After that, we create our widgets one by one. In this version we don't need to specify their positions. PyQt takes care to position them one after the other, vertically, since in line 27 we created a `QVBoxLayout()`, where the letter V stands for vertical.

The only thing that probably needs discussion in the code in Script 12.2 is that after we create each widget, we need to add it to our layout by calling the method `self.layout. addWidget(name_of_widget)`. We also connect certain actions on widgets to functions by calling the `connect(funcname)` method. Other than that, the function `set_wave- form()` is a bit different than the rest of the functions. In this function we iterate over all `QRa- dioButton()` objects and check which one is checked. When we find the one that is checked, we set the corresponding waveform to our oscillator, and break the loop.

In the main `if __name__ == "__main__"` chunk, we create a `QApplication()` object. Then we create an object of the class we defined and call its `show()` method, which it has inherited from its parent class. This method keeps our Script alive. The last like takes care to exit the program when we close its window. The rest of the code should be easy to understand by reading it. PyQt is a rather intuitive module, and its widgets look a bit fresher than tkinter's. The functionality of this GUI program is identical to the one from the previous section, even though it looks much different.

12.4 Conclusion

Pyo contains its own set of widgets that are functional, intuitive, and have a fresh look. With this set of GUIs, users can do almost anything that is needed when interacting with PyoObjects. Certain limitations, like radio buttons and numeric value entries might create the need to code our own GUI widgets. Another factor that can render Pyo's GUIs unavailable is the Python version Pyo is built against. It is possible that Pyo is used with a Python version that wxPython has no wheels for (yet), in which case it is not possible to use Pyo's GUIs. For these reasons we experimented with other Python GUI modules.

We have created two different GUIs to control the same attributes of a PyoObject in the exact same way. This way, this chapter highlights the fact that in computer programming there is not one way of doing things, but we can approach tasks from different points of view and end up with the same result. Each module we used has its advantages and disadvantages, compared to each other.

Tkinter is native to Python, so there is no need to install any additional software in our system. On the other hand, PyQt binds Python to a much more popular GUI toolkit than Tcl/Tk. PyQt might also be a bit more intuitive than tkinter, as creating widgets requires far less arguments. The GUIs programmed with this module also look fresher than the ones programmed with tkinter. Tkinter though, provides an easier way to place widgets at exact positions, whereas PyQt needs a bit more in-depth knowledge of the module to be able to have finer control over positioning the widgets.

In Chapter 6, we saw how we can use other languages or programs to communicate with Pyo over OSC. Instead of using Python modules for writing GUIs, we can use another language to do this, like openFrameworks, or Processing. Most programming languages targeting visuals, contain addons or libraries for GUIs. Being able to launch another program from the Python Script that runs Pyo, can make such a combination easy to use, as we don't need to launch different software separately, which can be more prone to error.

Finally, bear in mind that certain functionalities can be much harder to achieve with custom widgets, like making an oscilloscope, or a spectroscope. Pyo's GUIs cover these functionalities. Overall, you are encouraged to use Pyo's GUIs instead of writing your own, as this can be a cumbersome task that can shift your focus away from the creative process of creating sounds. Based on this last prompt, this chapter does not include exercises, as writing your own GUI is marginally within the scope of this book. Therefore, it is left to the discretion of the reader whether they will further develop the projects of this chapter, or whether they will write new ones from scratch.

Note

1 It is possible that PyQt6 depends on the OpenCV module for Python. If you get errors with this project, install this module as well.

13 Using Various Python APIs in Music Scripts

In this last chapter we will extend our Pythonic journey by including extra-musical elements to our scripts. We will utilise Python APIs for tasks that are not related to music, and see how we can use the information we get, in a musical context. This chapter aims also to highlight the vast amount of Python modules, for many tasks. We will use two APIs – the Google Maps API, and the Tweeter Streaming API. The former will provide us with latitude and longitude coordinates of randomly selected capital cities, and the latter will enable us to download tweets in real time.

For the first project, we will need an easy way to make a random selection from a list of capital cities, so we don't have to write long lists manually. For the second project, we will need to analyse the sentiment of the tweets, and a way to translate them, in case they are not in English. All this can be achieved in Python, as modules for each of these tasks are available.

Besides highlighting the flexibility of Python, this chapter aims to give inspiration for artistic projects that extend beyond pure sound, to a more conceptual realm. The two projects we will realise in this chapter function as proposals on how a sound artwork can be conceptually enriched. Contextualised by elements that initially seem to not relate to sound or music, in a way that eventually bridges the gap between the musical and the extra-musical, we will realise projects that use sound to project a critique on various matters.

What You Will Learn

After you have read this chapter you will:

- Know about the Google Maps and Twitter Streaming APIs
- Be able to retrieve geolocation information in Python
- Be able to stream tweets based on keyworks, in Python
- Know how to apply sentiment analysis on text

13.1 What Are Python APIs? .

As already mentioned in this book, an API is an abbreviation for Application Programming Interface. The two APIs we will use, provide an easy way to access information from their data bases. For example, with the Google Maps API, we can retrieve the latitude and longitude coordinates of a city as shown in Shell 13.1. The third and fourth lines are part of the interface

DOI: 10.4324/9781003386964-13

provided by the googlemaps Python module, that provide geolocation data of a city. Once we create a Google Maps `Client()` object, we can easily get the latitude and longitude values using this API.

Shell 13.1: Getting the latitude and longitude coordinates of Paris with the Google Maps API in Python

```
>>> import googlemaps
>>> gmaps = googlemaps.Client(key='your Google Maps API key')
>>> lat = gmaps.geocode("Paris")[0]['geometry']['location']['lat']
>>> lng = gmaps.geocode("Paris")[0]['geometry']['location']['lng']
>>> print(lat, lng)
48.856614 2.3522219
```

13.2   Using Google Maps API to Retrieve Geolocation Information as Musical Input

To be able to use the Google Maps API in Python, we must install the googlemaps module. To do this, type `pip install googlemaps` in a terminal. We will also need the pycountry and the countryinfo modules. Install them with pip, as you installed googlemaps. We must also get an API key from Google, to be able to make geolocation coordinates queries. To do this, you must create a project in Google, provide some information, and enable some features. You can find detailed information on the Google's developers documentation pages.[1]

Go ahead and create a project and enable the APIs as instructed, and then follow the "Using API Keys" button at the bottom of the documentation page, to create your key. Mind that these are the instructions at the time of writing. It is possible that in the future these steps will change. Nevertheless, it should be a rather easy process, as Google tries to make it intuitive so that developers can easily use their services.

13.2.1   *Contextualising Our Project*

Once you have your API key, you are ready to use the module and get geolocation coordinates. A question that might arise at this point is how can we use these coordinates in a musical context. One idea is to use these values as MIDI notes to control the frequencies of oscillators. Another idea is to create sequences with the Events Framework. What is important in an interdisciplinary artwork, is the context within which the various fields are inter-connected. Since we are dealing with geolocation data, we can approach our project from the point of view of localism, by associating sound to cities.

This association is possible by projecting the currently queried city and playing the sound that results from the mapping of the geolocation coordinates to MIDI notes or sequences. Such an association is likely to be random, but that can serve our concept well. We can assume that there are two notions of localism, a stereotypical one, and a subjective one. We can also assume that a stereotypical notion considers certain cities to be safe, beautiful or romantic, and others to be infamous and dangerous. A subjective notion of localism though can be the complete opposite, as it depends on personal experiences and the feeling of belonging. By making random associations between cities and sounds, we question this notion, as a stereotypically beautiful city might produce harsh and inharmonic sounds, and a stereotypically dangerous city might produce soft and harmonic sounds. On top of this, what sounds good – harmonic and soft, or inharmonic and harsh – is completely subjective.

Another aspect we can employ is rhythm, where randomly created rhythms, based on these geolocation data, can either produce steady or more broken beats. Whether one or the other rhythm sounds good is also subjective. By juxtaposing the harmonic element with the rhythmic one, we can reinforce our concept, as the aesthetics of the resulting sound will be more complex, as the matter of localism is. In this context, we parallel aesthetics – which are subjective – to the stereotypical notion of localism. Of course, this chapter does not aim to create a complete artwork, but only serves to provide inspiration and project the multitude of available Python modules that facilitate the creation of inter-disciplinary works.

13.2.2   Writing the Code

Let's now write the code for this project. Apart from the googlemaps, pycountry, and country-info modules, we will also need the python-osc module. Go ahead and install this one as well, if you did not do so in Chapter 6. We will also use the native multiprocessing module, but being native, we don't need to install it explicitly.

Ingredients:

- Python3
- The Pyo module
- The wxPython module, if you want to use Pyo's GUI
- The googlemaps module
- A Google Maps API key
- The pycountry module
- The countryinfo module
- The python-osc module
- A text editor (preferably an IDE with Python support)
- A terminal window in case your editor does not launch Python scripts

Process:

Let's break the code in smaller chunks, as we need to discuss some new concepts. Script 13.1 imports all the necessary modules and sets some constants and variables. The multiprocessing module is necessary so we can do the geolocation coordinates queries in a different thread than the one running our Python script, because getting these coordinates is a strenuous process, and if we run it in the main thread of our script, we will get drop outs (audible clicks) in our sound. The python-osc module is necessary so we can send data from the child thread to the main thread easily.

Script 13.1 Importing the modules and setting constants and variables.

```
1 from multiprocessing import Process
2 import googlemaps
3 from countryinfo import CountryInfo
4 import pycountry as cntry
5 from pythonosc.udp_client import SimpleUDPClient
6 import pyo
7 import random
8
```

```
 9 RAMPTIME = 5
10 ENVDUR = .125
11
12 ip= '127.0.0.1'
13 port = 5100
14 client = SimpleUDPClient(ip, port)
15
16 # the key argument below is left blank as this is private
17 gmaps = googlemaps.Client(key="")
18
19 num_countries = len(cntry.countries)
20
21 # minimum and maximum values of latitude and longitude
22 latminmax = [-85.05115, 85]
23 lngminmax = [-180, 180]
24
25
```

In lines 12 to 14 we set the various OSC parameters of our `SimpleUDPClient` object, and in line 17, we create our `googlemaps.Client()` object. The empty string should be filled with your personal Google Maps API key. In line 19 we get the number of countries available from the pycountry module. In lines 22 and 23 we store the minimum and maximum values of the latitude and longitude coordinates, so we can easily map them to the MIDI range later on.

Script 13.2 includes a single function that maps values from one range to another. The arguments to this function are the value to be mapped, the minimum value of the current range, the maximum value of the current range, and the minimum and maximum values of the desired range. This function will be used in a few places in our code. To scale a latitude value of 25.5 from its original range to the MIDI range, we need to call it like this: `mapped = mapval(25.5, latminmax[0], latminmax[1], 0, 127)`.

Script 13.2 Function that maps values from one range to another.
```
26 def mapval(val, inmin, inmax, outmin, outmax):
27     inspan = inmax - inmin
28     outspan = outmax - outmin
29     scaledval = float(val - inmin) / float(inspan)
30     return outmin + (scaledval * outspan)
31
32
```

The next script includes the function that makes the geolocation coordinate queries. This function will be called with a random number in the range of the total number of countries available by the pycountry module. We will try to get the capital city of the randomly chosen country, in lines 36 to 38. Some country indexes result in a `KeyError` when passed to the countryinfo module, so we remedy this with the `try/except` block and the `while` loop that tests if the `city` variable is None. When we exit the loop, we try to get the latitude and longitude values. Some cities return an empty string when passed to the `googlemaps.Client.geocode()` function, so we need this try/except block here as well, otherwise, if the geocode variable results in an empty list, trying to index it with 0 will fail. We then map the coordinates to the MIDI range, get the minimum value of the two as an integer, which we will use as our base MIDI note, and we get the absolute difference between the original coordinates, as an integer wrapped around 12, so it stays within a MIDI octave. We map this difference to a new range for the feedback of a `SineLoop()` oscillator we will use, and the number of taps we will pass to our rhythm generator, which we will see later.

Script 13.3 The geolocation coordinates query function.

```
33 def getloc_proc(ndx):
34     city = None
35     while city is None:
36         try:
37             country = list(cntry.countries)[ndx].name
38             city = CountryInfo(country).capital()
39         # some countries produce key errors
40         # because the function call above results to NoneType
41         except KeyError:
42             pass
43     try:
44         geocode = gmaps.geocode(city)
45         try:
46             lat = geocode[0]['geometry']['location']['lat']
47             lng = geocode[0]['geometry']['location']['lng']
48             # map the coordinates to the MIDI range
49             lat_mapped = mapval(lat, latminmax[0],
50                                 latminmax[1], 0, 127)
51             lng_mapped = mapval(lng, lngminmax[0],
52                                 lngminmax[1], 0, 127)
53             lat_lng_min = int(min(lat_mapped, lng_mapped))
54             # get the difference of the original coords
55             # wrapped around 12
56             diff = abs(int(lat) - int(lng)) % 12
57             feedback = mapval(diff, 0, 11, 0, 0.25)
58             taps = int(mapval(diff, 0, 11, 1, 24))
59             # write the string that will be printed
60             print_str = f"{city}: {lat}, {lng} " + \
61                         f"mapped to: {lat_lng_min} " + \
62                         f"{lat_lng_min+diff} " + \
63                         f"with feedback: {feedback} " + \
64                         f"and taps: {taps}"
65             print(print_str)
66             list_to_send = [lat_lng_min,
67                             lat_lng_min+diff,
68                             feedback, taps]
69             client.send_message("/latlng", list_to_send)
70         # some cities result in an empty list
71         # returned by gmaps.geocode(city)
72         except IndexError:
73             print(f"Index error with {city}")
74     except googlemaps.exceptions.HTTPError:
75         print("HTTP error")
76
77
```

Once we get all the values we need, we create a string to print on the console with the city, its latitude and longitude coordinates, the MIDI note values we will use, and the feedback amount and the number of taps. The two MIDI notes will be the minimum mapped coordinate, and this value added to the difference. This way, we will create simple chords that can sound either harmonic or inharmonic, depending on their relationship, which will be defined by the diff variable. Once we print this string on the console, we create a list with all the values we print and send it via OSC with our SimpleUDPClient().

If the queried city results in an empty string, and the try block in line 45 fails, we print this on the console, with the equivalent except, in line 72. It is also possible to get an HTTPError, raised by the googlemaps module. If this happens, we also print it on the console.

Script 13.4 includes the last two functions of our script. The get_latlng() function will be called by a Pattern() object. This function creates a Process object imported from the multiprocessing module. We pass the target function to its target kwarg, and a random value in the range of num_countries as the argument to this target function, passed as a one-value tuple, to the args kwarg. Once we create this object, we call its start() method. This will launch another Python process, running in a separate thread, so our main thread is not disturbed by it, and our sound is not affected.

Script 13.4 The get_latlng() and setvals() functions.
```
78 def get_latlng():
79     proc=Process(
80             target=getloc_proc,
81             args=(random.randrange(num_countries),)
82         )
83     proc.start()
84
85
86 def setvals(address, *args):
87     # args is a tuple, so we convert it to a list
88     midivals.setValue([args[0], args[1]])
89     feedback.setValue(args[2])
90     beat.setTaps(args[3])
91
92
```

The last function, setvals(), will be called by an OscDataReceive() object that will receive OSC messages from the child thread, with the list with the MIDI notes, the feedback value, and the number of taps. We will assign the first two values of the args tuple to midivals, which is a SigTo() object we will create in the next script, the third to feedback, which is again a SigTo() object, and the last to beat, which is a Euclide() object. Euclide() is a Pyo class that creates Euclidean rhythms.

Script 13.5 concludes our code for this project. In this script we create all the Pyo objects. The audio generators are two SineLoop() streams, created in line 100. The streams are two because the streams of the midivals object of the SigTo() class are two, since we pass a list with two values to its first argument. This two-stream SigTo() object is then fed to the freqs object of the MToF() class, that converts MIDI values to frequencies, which is set as the first argument to the SineLoop() object, resulting in the latter having two audio streams. The feedback object has one stream, as the feedback amount of both oscillator streams will be the same.

Script 13.5 The Pyo objects.

```
 93 s = pyo.Server().boot()
 94 midivals = pyo.SigTo([0, 0], time=RAMPTIME/2)
 95 feedback = pyo.SigTo(0, time=RAMPTIME/2)
 96 freqs = pyo.MToF(midivals)
 97 beat = pyo.Euclide(time=ENVDUR).play()
 98 tab=pyo.CosTable([(0,0), (64,1),(1024,1),(4096,0.5),(8191,0)])
 99 env = pyo.TrigEnv(beat, tab, dur=ENVDUR, mul=.5)
100 sines = pyo.SineLoop(freqs, feedback=feedback, mul=env).out()
101 oscrecv = pyo.OscDataReceive(5100, "/latlng", setvals)
102 pat = pyo.Pattern(get_latlng, time=RAMPTIME)
103 pat.play()
104 s.gui(locals())
```

In line 97, we create our Euclidean rhythm generator. We use its default kwargs and only set its `time` kwarg. Then we create an envelope with the `CosTable()` class, and a `TrigEnv()` object, to play this envelope, triggered by our Euclidean rhythm generator. In line 101 we create the `OscDataReceive()` object that will receive OSC messages from the child thread of this program, and we set the `setvals()` function from Script 13.4 to its last argument.

Finally, we create a `Pattern()` object that will call the `pat_latlng()` function every RAMPTIME seconds, we start it, and we launch Pyo's GUI so we can start the Pyo server. Once we click on the "Start" button of the server window, we will start hearing sound played in rhythmic patterns, and we will see information like the line below, printed on our console.

```
Athens: 37.9838096, 23.7275388 mapped to: 71 73 with feedback:
0.045454545454545456 and taps: 5
```

13.2.2.1   Remedy for Excessive CPU Usage

There is a small caveat with this code. The `KeyError` that might occur and will be `excepted` in line 41 in Script 13.3, can cause our Python script to consume almost 100% of the CPU of the core it is running in. This will probably not affect our sound, as the core, or thread of this function will be different than the one that generates the sound. But as soon as we click on the "Stop" button on Pyo's server window, and try to quit the program by clicking "Quit", it is very likely that our program will stall, and we will have to hit Ctl+C to forcefully quit it (or use `killall` on Unix systems).

We can avoid this, but we need to take a few extra steps. First, we will need to exclude the countries that produce this `KeyError` from the list of country indexes. To do this, we will write a script that includes the `try/except` block of Script 13.3 that `excepts` this `KeyError`. The code is shown in Script 13.6. All we do is create a file and write the indexes of cities that produce this error. We do this by writing the index of the for loop to the file, only when this error is raised. When the loop is done, we close the file and our script ends.

Script 13.6 Script to store the indexes of countries that produce a KeyError.

```
1 from countryinfo import CountryInfo
2 import pycountry as cntry
3
4 num_countries = len(cntry.countries)
5
```

```
6 def exclude_key_error_countries():
7       f = open("excluded_countries.txt", "w")
8       for i in range(num_countries):
9           try:
10              CountryInfo(
11                  list(cntry.countries)[i].name
12              ).capital()
13          except KeyError:
14              f.write(str(i)+"\n")
15      f.close()
16
17 if __name__ == "__main__":
18     exclude_key_error_countries()
```

If you run this script, you will end up with a file called *excluded_countries.txt*. Make sure you run Script 13.6 from the directory of your Google Maps API script, so the latter can spot the file created by the former. To use this file, we will need to make a few changes to the original script. Script 13.7 shows the code that stores only countries that don't raise this error in a list. Go ahead and insert this code to the main script, right after all the code in Script 13.1.

Script 13.7 Store indexes of countries that don't raise a KeyError.
```
26 lcountry = []
27 # exclude countries that raise a KeyError
28 excluded_countries = []
29 # the indexes of these countries have been stored
30 # in a text file
31 f = open("excluded_countries.txt", "r")
32 for line in f:
33     line = line.rstrip()
34     excluded_countries.append(int(line))
35 f.close()
36 # list only countries that don't raise a KeyError
37 for i in range(num_countries):
38     if i not in excluded_countries:
39         lcountry.append(i)
40
41
```

Three more minor changes need to be made in the main script. In Script 13.3, in line 37 – which should have a different line number now – change ndx, passed as a list index, to lcountry[ndx], as we will now be looking at a list with indexes, and not count on the number of countries available from the pycountry module. We can also get rid of the try/except block that includes the country and capital city choice, as we don't run the risk of a KeyError anymore. The last change we need to make is in Script 13.4. Change the num_countries variable in line 81 – which should also be different now – to len(lcountry), as we now need to pass a random value in the range of the length of the lcountry list that contains the indexes of the countries that don't raise a KeyError.

  If you apply these changes and run the script again, you will see that the CPU used is far less than what was used with the initial version of the code, and when you want to quit the program, it quits without hanging. The resulting program queries coordinates of randomly selected capital cities, and produces random harmonies, textures, and rhythms, based on these coordinates. The association made between cities and sound is random and projects the subjective notion of localism, in contrast to a stereotypical notion of it.

Some cities produce harsher sounds that others, and some cities produce more steady rhythms than others. Some cities produce very similar results with other cities, as the difference between their latitude and longitude is the same, resulting in the same note relationships, feedback values, and number of taps. Within the concept of questioning the notion of localism, this serves us well, as certain cities are rendered equal, even if they might be considered to be completely different. The full code of this project is shown in Script 13.8.

Script 13.8 Full code for the Google Maps project.

```
 1 from multiprocessing import Process
 2 import googlemaps
 3 from countryinfo import CountryInfo
 4 import pycountry as cntry
 5 from pythonosc.udp_client import SimpleUDPClient
 6 import pyo
 7 import random
 8
 9 RAMPTIME = 5
10 ENVDUR = .125
11
12 ip= '127.0.0.1'
13 port = 5100
14 client = SimpleUDPClient(ip, port)
15
16 # the key argument below is left blank as this is private
17 gmaps = googlemaps.Client(key='')
18
19 num_countries = len(cntry.countries)
20
21 # minimum and maximum values of latitude and longitude
22 latminmax = [-85.05115, 85]
23 lngminmax = [-180, 180]
24
25
26 lcountry = []
27 # exclude countries that raise a KeyError
28 excluded_countries = []
29 # the indexes of these countries have been stored
30 # in a text file
31 f = open("excluded_countries.txt", "r")
32 for line in f:
33     line = line.rstrip()
34     excluded_countries.append(int(line))
35 f.close()
36 # list only countries that don't raise a KeyError
37 for i in range(num_countries):
38     if i not in excluded_countries:
39         lcountry.append(i)
40
```

```
41
42 def mapval(val, inmin, inmax, outmin, outmax):
43     inspan = inmax - inmin
44     outspan = outmax - outmin
45     scaledval = float(val - inmin)/float(inspan)
46     return outmin + (scaledval * outspan)
47
48
49 def getloc_proc(ndx):
50     city = None
51     while city is None:
52         try:
53             country=list(cntry.countries)[lcountry[ndx]].name
54             city = CountryInfo(country).capital()
55         # some countries produce key errors
56         # because the function call above results to NoneType
57         except KeyError:
58             pass
59     try:
60         geocode = gmaps.geocode(city)
61         try:
62             lat = geocode[0]['geometry']['location']['lat']
63             lng = geocode[0]['geometry']['location']['lng']
64             # map the coordinates to the MIDI range
65             lat_mapped = mapval(lat, latminmax[0],
66                             latminmax[1], 0, 127)
67             lng_mapped = mapval(lng, lngminmax[0],
68                             lngminmax[1], 0, 127)
69             lat_lng_min = int(min(lat_mapped, lng_mapped))
70             # get the difference of the original coords
71             # wrapped around 12
72             diff = abs(int(lat) - int(lng)) % 12
73             feedback = mapval(diff, 0, 11, 0, 0.25)
74             taps = int(mapval(diff,0, 11, 1, 24))
75             # write the string that will be printed
76             print_str = f"{city}: {lat}, {lng} " + \
77                         f"mapped to: {lat_lng_min} " + \
78                         f"{lat_lng_min+diff} " + \
79                         f"with feedback: {feedback} " + \
80                         f"and taps: {taps}"
81             print(print_str)
82             list_to_send = [lat_lng_min,
83                             lat_lng_min+diff,
84                             feedback, taps]
85             client.send_message("/latlng", list_to_send)
86         # some cities result in an empty list
87         # returned by gmaps.geocode(city)
88         except IndexError:
89             print(f"Index error with {city}")
90     except googlemaps.exceptions.HTTPError:
91         print("HTTP error")
92
```

```
93
94  def get_latlng():
95      proc=Process(
96              target=getloc_proc,
97              args=(random.randrange(len(lcountry)),)
98          )
99      proc.start()
100
101
102 def setvals(address, *args):
103     # args is a tuple, so we convert it to a list
104     midivals.setValue([args[0], args[1]])
105     feedback.setValue(args[2])
106     beat.setTaps(args[3])
107
108
109 s = pyo.Server().boot()
110 midivals = pyo.SigTo([0, 0], time=RAMPTIME/2)
111 feedback = pyo.SigTo(0, time=RAMPTIME/2)
112 freqs = pyo.MToF(midivals)
113 beat = pyo.Euclide(time=ENVDUR).play()
114 tab=pyo.CosTable([(0,0),(64,1),(1024,1),(4096,0.5),(8191,0)])
115 env = pyo.TrigEnv(beat, tab, dur=ENVDUR, mul=.5)
116 sines = pyo.SineLoop(freqs,feedback=feedback,mul=env).out()
117 oscrecv = pyo.OscDataReceive(5100, "/latlng", setvals)
118 pat = pyo.Pattern(get_latlng, time=RAMPTIME)
119 pat.play()
120 s.gui(locals())
```

Documentation Page:

`Euclide()`: https://belangeo.github.io/pyo/api/classes/triggers.html#euclide

13.3 Mining Tweets from Twitter and Applying Sentiment Analysis to Control Sound

For our next project, we will download tweets from Twitter in real-time, using the Twitter Streaming API. This API gives us access to the latest tweets, based on hashtags we provide. To get access to this API, we need to sign up for a Twitter developer account, and get some keys and tokens from Twitter. The documentation of the developer platform of Twitter[2] should get you up and running. Go through the procedure in the documentation, and get a Consumer Key and Consumer Secret from the *API Key and Secret* link, and an Access Token and an Access Token Secret, which you can get from the *Access Token and Secret* link. We will also need a Bearer Token. Twitter's documentation[3] on how to get one should be helpful enough. Make sure you have these keys and tokens stored somewhere safe, and then install the Twitter Streaming API module with `pip install tweepy`. Note that the keys and tokens are private, and you should not share them with anyone.

For this project, we will apply sentiment analysis on the tweets we will be downloading. This will be achieved with the vaderSentiment module. Go ahead and install it with pip. Mind though that this module is not capable of analysing the sentiment of all languages. For this reason, we will need to translate tweets in case these are not in English. We will use the Google Translate service from the Google Cloud Platform. Go through the procedure specified in the documentation pages of this service.[4] Mind that the 2.0.1 version is sufficient for our project, so follow the instructions up to *Cloud Translation – Basic client libraries*.

13.3.1 Contextualising Our Project

As with the previous project, in an artistic context, it is somewhat necessary to contextualise our idea. Since we will be downloading tweets which we will analyse based on their sentiment, we can use sound to project this sentiment. The output of the sentiment analysis comes as a Python dictionary with the following keys: `'neg'`, `'neu'`, `'pos'`, and `'compound'`. These keys stand for "negative", "neutral", "positive", and obviously "compound". The values of the first three keys are in the range between 0 and 1, and for the `'compound'` key, the values are in the range between -1 and 1, where negative compound values denote a compound sentiment leaning toward negative, and vice versa.

We can parallel negative sentiments with an inharmonic and harsh sound, neutral with a bright, harmonic, but neither major nor minor chord, and positive with a soft, major chord. For this project we can omit the compound sentiment value because we will be mixing the sounds for the other three sentiment keys, so this mix will serve as the sound for the compound sentiment. The amplitude of each sound in the mix will be controlled by the respective sentiment value, so, depending on the overall sentiment, each sound will prevail accordingly.

13.3.2 Writing the Code

Now that we have all the necessary modules and we have contextualised our idea, we can start writing the code. As with the previous project, we will break it up in smaller chunks, so we can focus our discussion on each new element separately.

Ingredients:

- Python3
- The Pyo module
- The tweepy module
- A Twitter Consumer Key, Consumer Secret, Access Token, Access Token Secret, and Bearer Key
- The google-cloud-translate==2.0.1 module
- A Google service key
- The vaderSentiment module
- A text editor (preferably an IDE with Python support)
- A terminal window in case your editor does not launch Python scripts

Process:

Script 13.9 shows the code for importing all the necessary modules and all the Pyo classes. For the negative sentiment, we create a `SuperSaw()` oscillator. This is a series of band-limited sawtooth oscillators where we can control how much detuned they will be from one another. We set a low frequency, a high detune value, and its balance set toward its detuned oscillators. For the neutral sentiment, we create a G suspended fourth chord with a three-stream `SineLoop()`

oscillator, setting its `feedback` to 0.1, so we get some brilliance. We use a suspended fourth chord so it is neither major nor minor. Finally, for the positive sentiment we create a D major chord, starting from the D of the previous chord, with pure sines, so we get a soft sound.

Script 13.9 Importing the modules and creating the Pyo classes.

```
 1 import tweepy
 2 from vaderSentiment.vaderSentiment \
 3 import SentimentIntensityAnalyzer
 4 from twitter_api_credentials import *
 5 from google.cloud import translate_v2 as translate
 6 import pyo
 7
 8 RAMPTIME = .5
 9
10 s = pyo.Server().boot()
11 s.start()
12
13 neg_amp = pyo.SigTo(0, time=RAMPTIME)
14 neg = pyo.SuperSaw(freq=50, detune=0.8, bal=0.9, mul=neg_amp)
15 neu_amp = pyo.SigTo(0, time=RAMPTIME)
16 neu = pyo.SineLoop(freq=pyo.MToF(pyo.Sig([55,60,62])),
17 feedback=.1, mul=neu_amp)
18 pos_amp = pyo.SigTo(0, time=RAMPTIME)
19 pos = pyo.Sine(freq=pyo.MToF(pyo.Sig([62,66,69])),
20                 mul=pos_amp)
21
22 mixer = pyo.Mixer(outs=1, chnls=7)
23 mix = pyo.Mix(mixer, voices=2, mul=.5).out()
24 # add inputs and set amps for the mixer
25 mixer.addInput(0, neg)
26 mixer.setAmp(0, 0, .133)
27 for i in range(3):
28     mixer.addInput(i+1, neu[i])
29     mixer.addInput(i+4, pos[i])
30     mixer.setAmp(i+1, 0, .133)
31     mixer.setAmp(i+4, 0, .133)
32
33
```

We mix all these signals with a `Mixer()` object with one output, and set a low amplitude to each input channel of the mixer, so we don't get a loud and distorted sound. Note that we add each stream of the three-stream oscillators to the mixer separately, so we can send all signals to both channels of a stereo setup. If we passed the three-stream oscillators as one object to a single input channel of the mixer, the streams would be routed to alternating channels, ending up in the first and last notes of each stream sounding from the left speaker, and the middle note sounding from the right speaker. The `SuperSaw()` oscillator doesn't face this issue, and we add it to the mixer as is, since the actual object is a single audio stream.

Apart from the necessary modules that we import in our script, we also import all the contents of a file called *twitter_api_credentials*, in line 4. This is a Python file that we will create later, where we will store the keys and tokens we got from Twitter, so we can authenticate our developer app, and get access to the Twitter Streaming API.

Script 13.10 shows the code for authenticating to Twitter, creating a `tweepy.API()` object, and an object of the class we will define to download tweets in real-time. The keys and tokens passed to the `tweepy.OAuth1UserHandler()` object in lines 77 to 80 will be included in the *twitter_api_credentials.py* file that was mentioned above. So is the `BEARER_TOKEN`, passed to our `TweetDowloader()` class, in line 94. This custom class will take this token, together with the `tweepy.API()` object as arguments, as we see in this line.

Script 13.10 Authenticating to Twitter, and creating the API() and the TweetDowloader() objects.

```
75  if __name__ == "__main__":
76      # Authenticate to Twitter
77      auth = tweepy.OAuth1UserHandler(CONSUMER_KEY,
78                                      CONSUMER_SECRET,
79                                      ACCESS_TOKEN,
80                                      ACCESS_TOKEN_SECRET)
81
82      # Create Twitter API object
83      api = tweepy.API(auth)
84
85      # make sure the app's credentials are correct
86      # otherwise exit
87      try:
88          api.verify_credentials()
89          print("Authentication OK")
90      except:
91          print("Error during authentication")
92          exit()
93
94      tweets_client = TweetDownloader(BEARER_TOKEN, api)
95      # delete previous rules so we can set new ones anytime
96      current_rules = tweets_client.get_rules().data
97      for rule in current_rules:
98          tweets_client.delete_rules(rule.id)
99      # set whatever keywords (including hashtags)
100     # you want to download
101     tweets_client.add_rules(
102         [tweepy.StreamRule(value="#a_hashtag")]
103     )
104     tweets_client.filter()
```

To download tweets based on keywords, we have to set some rules to our streaming object. These rules persist even after we quit the script, so before setting a rule, it is a good idea to delete previous rules. If we don't delete these previous rules, when we add a new rule, it will just be added to the previous ones. If we want to run this script and download tweets based on a new hashtag only, we need to delete the previous rules. This is done in lines 96 to 98.

Once we delete the old rules, we can add a new rule, which happens in lines 101 to 103. Replace the string in this line with any hashtag or keyword you like. It is possible to set more than one keywords, by adding a `tweepy.StreamRule()` object for each keyword in the list in line 102. After we add our new rules, we need to call the `filter()` method of our class. All these methods of our class that we call in Script 13.10 are inherited from a parent class. The code of our class is shown in Script 13.11.

Script 13.11 The tweet downloading class.

```
34  class TweetDownloader(tweepy.StreamingClient):
35      def __init__(self, bearer_token, api, **kwargs):
36          self._api = api
37          self.sent = SentimentIntensityAnalyzer()
38          self.translator = translate.Client()
39          super().__init__(bearer_token, **kwargs)
40
41      def on_tweet(self, tweet):
42          is_retweet = False
43          is_translated = False
44          status = self._api.get_status(tweet.id,
45                                            tweet_mode="extended")
46          try: # if it is a retweet
47              full_tweet = status.retweeted_status.full_text
48              is_retweet = True
49          except AttributeError: # otherwise it is a tweet
50              full_tweet = status.full_text
51          if status.lang != "en":
52              translator = self.translator.translate(
53                              full_tweet,
54                              target_language="en"
55                          )
56              sent = self.sent.polarity_scores(
57                          translator['translatedText']
58                      )
59              is_translated = True
60          else:
61              sent = self.sent.polarity_scores(full_tweet)
62          if is_retweet:
63              print(f"retweet: {full_tweet}")
64          else:
65              print(f"tweet: {full_tweet}")
66          if is_translated:
67              print(f"translated sentiment: {sent}\n")
68          else:
69              print(f"sentiment: {sent}\n")
70          neg_amp.setValue(sent['neg'])
71          neu_amp.setValue(sent['neu'])
72          pos_amp.setValue(sent['pos'])
73
74
```

Our class inherits from the `tweepy.StreamingClient()` class, and takes the bearer token and the `tweepy.API()` object as arguments, together with any possible additional kwargs. In its `__init__()` method, we store the `tweepy.API()` object to an instance variable, and create a `SentimentIntensityAnalyzer()` object and a translation client. We then call the `__init__()` method of the super class with the same arguments as our child class, except for the `tweepy.API()` object.

All the tweet processing happens in the `on_tweet()` method. This method is defined in the parent class, but it just `passes`, so if we want to do anything with it, we need to override it, and write code that fits our needs. In this overriden method, we create two Booleans to determine whether a tweet is a retweet or not, and whether it has been translated or not. Then we set the mode of the API to "extended" so we can download the full text of the tweet, otherwise we will be downloading a compacted version of it. We achieve this by calling the `get_status()` method of the `tweepy.API()` object, where we pass the `id` attribute of an object of the class `tweepy.tweet.Tweet()` that is passed to this method (the `tweet` object), and setting the `tweet_mode` kwarg to "extended". We then `try` to get the text of a retweet, and if that fails, we get the text of a tweet.

In line 51 we check if the tweet is not in English, and if this is `True`, we translate it using our Google translation client. Once the tweet has been translated, we get its sentiment in lines 56 to 58, and set the `is_translated` Boolean to `True`, so we can print correct information later on. If the tweet is in English, no translation is needed and we get the sentiment straight from the `full_tweet` string. When we are done with translating and getting the sentiment, we print the tweet, stating if it is a retweet or a tweet, and the sentiment, stating whether it is retrieved from a translated tweet or from the original one. Then we pass the sentiment values to the corresponding `SigTo()` objects that control the amplitude of the three sound sources.

Before we are able to run the script, we need to write the Python file that will contain the keys and tokens that we need. This is shown in Script 13.12. The strings in this script are left blank, because these keys and tokens are private. You should not share your keys and tokens with anyone. Make sure you store them in a safe place. If you ever want to share your scripts that use the Twitter API, make sure to delete the strings that contain them first, and hand in your code with empty placeholders, where each user can put their own information, just like it is done in Script 13.12. Mind also that it is possible that these tokens have a certain time duration within which they are valid. If they are to be invalidated, you will be notified by Twitter, most likely via email.

Script 13.12 The Twitter keys and tokens.
```
1 ACCESS_TOKEN = ""
2 ACCESS_TOKEN_SECRET = ""
3 CONSUMER_KEY = ""
4 CONSUMER_SECRET = ""
5 BEARER_TOKEN = ""
```

When you run the main script, if all has gone well with the authentication process, you should get an "Authentication OK" message printed, and after a while, depending on the keywords you use and how much traffic there is on Twitter with these keywords, you should start seeing tweets being printed, and the sound sources changing in volume. As with the previous project, and most projects in this book, the sound is rather simplistic, because we had to focus our discussion on the Twitter API side, and not on audio synthesis. You are encouraged to take this project further by refining the sound and the context. The full code is shown in Script 13.13.

Script 13.13 Full code for the Twitter project.
```
1 import tweepy
2 from vaderSentiment.vaderSentiment \
3 import SentimentIntensityAnalyzer
4 from twitter_api_credentials import *
5 from google.cloud import translate_v2 as translate
6 import pyo
7
8 RAMPTIME = .5
```

```
 9
10 s = pyo.Server().boot()
11 s.start()
12
13 neg_amp = pyo.SigTo(0, time=RAMPTIME)
14 neg = pyo.SuperSaw(freq=50, detune=0.8, bal=0.9, mul=neg_amp)
15 neu_amp = pyo.SigTo(0, time=RAMPTIME)
16 neu = pyo.SineLoop(freq=pyo.MToF(pyo.Sig([55,60,62])),
17 feedback=.1, mul=neu_amp)
18 pos_amp = pyo.SigTo(0, time=RAMPTIME)
19 pos = pyo.Sine(freq=pyo.MToF(pyo.Sig([62,66,69])),
20                   mul=pos_amp)
21
22 mixer = pyo.Mixer(outs=1, chnls=7)
23 mix = pyo.Mix(mixer, voices=2, mul=.5).out()
24 # add inputs and set amps for the mixer
25 mixer.addInput(0, neg)
26 mixer.setAmp(0, 0, .133)
27 for i in range(3):
28     mixer.addInput(i+1, neu[i])
29     mixer.addInput(i+4, pos[i])
30     mixer.setAmp(i+1, 0, .133)
31     mixer.setAmp(i+4, 0, .133)
32
33
34 class TweetDownloader(tweepy.StreamingClient):
35     def __init__(self, bearer_token, api, **kwargs):
36         self._api = api
37         self.sent = SentimentIntensityAnalyzer()
38         self.translator = translate.Client()
39         super().__init__(bearer_token, **kwargs)
40
41     def on_tweet(self, tweet):
42         is_retweet = False
43         is_translated = False
44         status = self._api.get_status(tweet.id,
45                                     tweet_mode="extended")
46         try: # if it is a retweet
47             full_tweet = status.retweeted_status.full_text
48             is_retweet = True
49         except AttributeError: # otherwise it is a tweet
50             full_tweet = status.full_text
51         if status.lang != "en":
52             translator = self.translator.translate(
53                             full_tweet,
54                             target_language="en"
55                         )
56             sent = self.sent.polarity_scores(
57                         translator['translatedText']
58                     )
59             is_translated = True
60         else:
```

```
61                     sent = self.sent.polarity_scores(full_tweet)
62              if is_retweet:
63                  print(f"retweet: {full_tweet}")
64              else:
65                  print(f"tweet: {full_tweet}")
66              if is_translated:
67                  print(f"translated sentiment: {sent}\n")
68              else:
69                  print(f"sentiment: {sent}\n")
70              neg_amp.setValue(sent['neg'])
71              neu_amp.setValue(sent['neu'])
72              pos_amp.setValue(sent['pos'])
73
74
75  if __name__ == "__main__":
76      # Authenticate to Twitter
77      auth = tweepy.OAuth1UserHandler(CONSUMER_KEY,
78                                      CONSUMER_SECRET,
79                                      ACCESS_TOKEN,
80                                      ACCESS_TOKEN_SECRET)
81
82      # Create Twitter API object
83      api = tweepy.API(auth)
84
85      # make sure the app's credentials are correct
86      # otherwise exit
87      try:
88          api.verify_credentials()
89          print("Authentication OK")
90      except:
91          print("Error during authentication")
92          exit()
93
94      tweets_client = TweetDownloader(BEARER_TOKEN, api)
95      # delete previous rules so we can set new ones anytime
96      current_rules = tweets_client.get_rules().data
97      for rule in current_rules:
98          tweets_client.delete_rules(rule.id)
99      # set whatever keywords (including hashtags)
100     # you want to download
101     tweets_client.add_rules(
102         [tweepy.StreamRule(value="#a_hashtag")]
103     )
104     tweets_client.filter()
```

Documentation Page:

SuperSaw(): https://belangeo.github.io/pyo/api/classes/generators.html#supersaw

13.4 Conclusion

We have seen how we can use Python APIs in our scripts, and how to use extra-musical information in a music context. Before we got into writing the code for our projects, we went through

the process of contextualising our concepts, something that is very important when creating inter-disciplinary artworks. The aim of both projects of this chapter was to use sound to project the concept. In the first project, we used sound to make a critique on the notion of localism, and how a stereotypical perception of localism can contradict a subjective one. We realised this critique by associating cities to sounds, based on mapping the geolocation coordinates to MIDI notes, and their relationship to other sound attributes.

The second project applied sentiment analysis on tweets downloaded from Twitter in real-time, using the Twitter Streaming API for Python. The sentiment analysis controlled the amplitude of three different sound sources, where each represented one of the three sentiments: negative, neutral, and positive. The overall mix of sounds served as a compound audio texture that aimed to reflect the overall sentiment of each tweet. Even though aesthetics are subjective, we used a more objective approach to sound, where a harsh and inharmonic sound was used for the negative sentiment, a harmonic, bright, but neither major nor minor chord was used for neutral, and a major and soft chord was used for the positive sentiment.

The sounds used to realise the two projects of this chapter were quite simple, as the focus of the discussion was on the APIs used, and how these can be integrated to the sound. These projects aimed to serve as an inspiration for the realisation of inter-disciplinary works, where sound is the main tool. They also aimed to demonstrate how various Python APIs work, and to project the multitude of available Python modules, for many tasks. You are encouraged to take these projects further, or to create your own, with different elements, even if these are not related to music or sound at all.

As a last note, both projects can be extended and include a visual aspect, where in the case of the first project, the queried city will be projected – this could also include the geolocation coordinates – and in the case of the second, the tweets will be projected. This can be achieved with openFrameworks, or another programming environment for visuals that supports the OSC communication protocol. It is beyond the scope of this book to show how this can be done in OF, but referring back to Chapter 6, you can get a starting point as to how to create an OF app that receives OSC. What is left, is to figure out how to display strings received via OSC.

13.5   Exercises

13.5.1   *Exercise 1*

Run the `getloc_proc()` function of the first project in the main thread of the program, to see how the strenuous process of getting the geolocation coordinates can affect the sound.

Tip: Don't use the multiprocessing module at all, just call the function straight from the `Pattern()` object. Mind to not forget the argument to this function.

13.5.2   *Exercise 2*

Experiment with different ranges for the feedback and the number of taps in the Google Maps project.

13.5.3   *Exercise 3*

Use the geolocation coordinate data to control another audio synthesis technique, like FM, or PM (Phase Modulation).

13.5.4   *Exercise 4*

Get tweets mentioning a specific user, instead of filtering according to keywords. This could be a politician, or a celebrity, that people tweet a lot about. Apply sentiment analysis the same way we have done in the second project.

Tip: Check the examples provided by the Twitter developer documentation.[5]

13.5.5   *Exercise 5*

Feed the sentiment analysis output to a Neural Network, that will control the parameters of a sound. Try to get sounds that project the sentiment as close as possible, based on your perception.

Tip: Revisit Chapter 7 and copy the necessary bits from the regression NN. Create a training dataset with some representative sets of sentiment values for all four sentiments. Augment your data with small variations on the input. Make sure you feed your network with a dataset of around 1,000 samples after applying augmentation.

Notes

1 https://developers.google.com/maps/documentation/geocoding/cloud-setup
2 https://developer.twitter.com/en/docs/twitter-api/getting-started/getting-access-to-the-twitter-api
3 https://developer.twitter.com/en/docs/authentication/oauth-2-0/bearer-tokens
4 https://cloud.google.com/translate/docs/setup
5 https://docs.tweepy.org/en/stable/examples.html

Conclusion

This book has covered a lot of ground, from the very basics of Python itself, through the very basics of digital audio in general, and audio with the Pyo module, to more advanced topics, including physical computing, network communication, and even AI. Throughout this book, very little use of languages other than Python has been made. This proves that Python is a very effective language, for many tasks, whether musical or not.

Being a general-purpose language, one might think that Python is not the best choice when it comes to audio, with the strict temporal constraints the latter entails. Pyo, though, is a domain-specific module, written in C – like most audio programming environments are – with its interface being pure Python. A domain-specific module for a general-purpose language provides a unique combination that can leverage tools for various projects, from the simplest to the most demanding.

This book aims to cover a wide range of audio and music-centred applications, nevertheless, in a generic context. When examples tend to become specific, like the Theremin simulation, or the use of the Google Maps and Twitter APIs, the aim is still to provide examples that can either fit various projects, or that can be easily modified to the needs of each user. Regardless of the scope of each example, specific or generic, state-of-the-art technology is used throughout the book, in a sonic context.

After the basics are covered, each chapter wanders into a musical or extra-musical field, where music starts to become a reality. This book is not only a handbook on digital audio synthesis techniques and their theory, but a source of knowledge and inspiration on how the computer can be used to create music. It is a hands-on cookbook, where the entire menu is music. As a reader, you are encouraged to take the examples of this book further and to explore the creative aspect of coding in general, more specifically using Python. Having read this book, you should be ready to take your own route and build your own artistic and musical vocabulary, through the use of this very popular language, and this very intuitive and efficient DSP module.

I hope this book will fill the gap created by the lack of literature on audio with Python, and find its place alongside books and manuals on other audio programming environments. Complete novices, but also beginner and advanced programmers, should equally find this book useful, in a sonic/Pythonic context. Even though this book covers a lot of ground, the reader is encouraged to check the documentation of Pyo itself,[1] as this includes numerous examples and ideas on how to create audio and music with this module.

Note

1 https://belangeo.github.io/pyo/index.html

DOI: 10.4324/9781003386964-14

Index